BUILD YOUR OWN
TELESCOPE

BUILD YOUR OWN
TELESCOPE

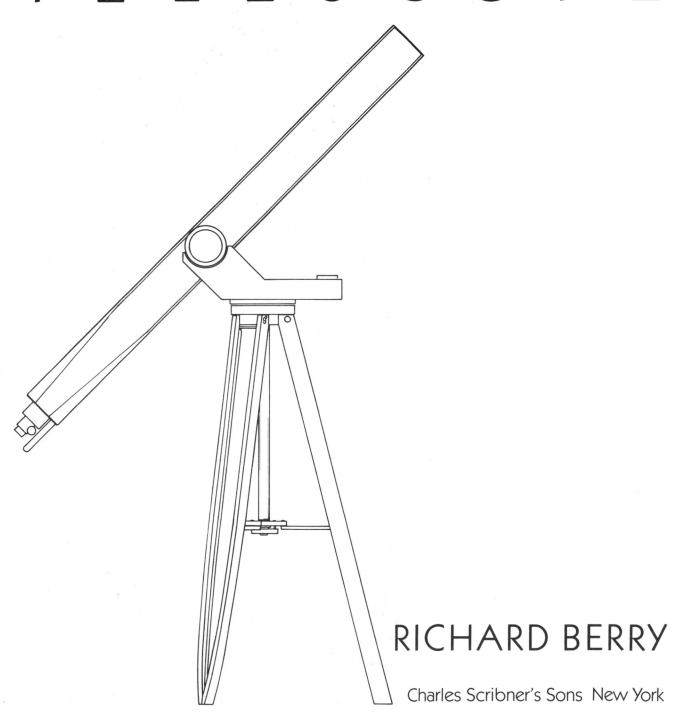

RICHARD BERRY

Charles Scribner's Sons New York

FOR MY PARENTS,
who for many years put up with things
that went bump in the night

Copyright © 1985 Richard Berry

Library of Congress Cataloging-in-Publication Data

Berry, Richard, 1946-
    Build your own telescope.

    Bibliography: p.
    Includes index.
    1. Telescopes.   2. Do-it-yourself work.
I. Title.
QB88.B47   1985   681'.412   85-18266
ISBN 0-684-18476-1

Published simultaneously in Canada by Collier Macmillan Canada, Inc.
Copyright under the Berne Convention.

1 3 5 7 9 11 13 15 17 19   H/C   20 18 16 14 12 10 8 6 4 2

Printed in the United States of America.

Contents

Introduction

This book was inspired, in large part, by requests from the readers of magazines I edit for plans for building telescopes. Often I was forced to refer readers to books I knew would not help them. For one thing, those books don't have plans, but only general guidelines. And they assume that if you want to build a telescope, you have a machine shop in your basement—and the skills to use it.

Like you, I am a do-it-yourselfer. When it comes to carpentry and metalworking, I am self-taught. Like me, I assume that you don't have a lot of time, money, or a well-equipped workshop full of tools to build a telescope with, and you don't have the time or inclination to chase all over the country after hard-to-find special parts. And I also assume that you'll want to use your telescope when you finish it.

I have designed each telescope in this book around materials and tools you should be able to find in any hardware or home-improvement store without spending an unreasonable amount of money. For the most part, the telescopes are made of wood and are without machined parts. I've tried to avoid designs that require expensive tools, or skills that require time and practice to acquire.

Like you, if I'm going to invest a significant amount of time and money in a project, I want the finished product to look good and work well. Consequently, I have tried to make the plans and instructions in this book as clear as possible; and to check the instructions, I've built and observed with every telescope described in it.

What does it take to build a telescope? Patience and a thoughtful approach are essential to do the job right. For the mechanical parts, you'll need a place where you can make more than a little sawdust and where no one will object to drilling, pounding, or the smell of fresh paint. A basement or garage is perfect. You'll need basic handtools—a drill, a hammer, a hand saw, a plane, screwdrivers, a putty knife, an open-end adjustable wrench—that type of thing. Power tools are handy but not essential.

For optical work, you'll need a double dose of patience. If you know people who have done optical work, seek their advice. You'll need a clean place to work in, a place with a fairly uniform temperature, free of moving dust and grit, preferably close to running water. The laundry room, a quiet corner of the basement, even the kitchen can serve. You'll need a sturdy table, some plastic pails, two glass disks, pitch, some little cans of abrasives and polishers, and about ten hours of working time per inch of aperture.

Where do you find specialized telescope supplies? See Appendix B. Some components—eyepieces, focusers, mirror cells, finder telescopes—you can make yourself, others you can buy from mail-order suppliers. You can purchase finished lenses and mirrors as well as supplies with which to make your own. If you want to know more about telescopes, optics, or telescopic observing, Appendix C lists books, catalogs, star maps, articles, and other sources of information.

Chapters 1, 2, and 3 cover things you should know about telescopes. Chapters 4 through 8 detail specific telescope projects, complete with plans; and chapters 9 through 11 cover telescope parts and optical work. Chapter 12 offers a prospectus on what you can see and do with your finished telescope. Follow the plans exactly if you wish, or improvise freely. After all, it's going to be *your* telescope!

RICHARD BERRY
Milwaukee, Wisconsin

BUILD YOUR OWN
TELESCOPE

Which Telescope is for You?

Everyone who has looked at the sky on a clear night has probably, at least once, wished for a telescope to better see the stars. Yet few people ever own one—they think a telescope is too complicated or too expensive, and the magical desire to see more of the heavens gets pushed aside or forgotten. Yet a powerful telescope capable of revealing the craters on the moon, the rings of Saturn, and the distant pinwheels of galaxies is neither too difficult to use nor too complex to build at home.

You can build most of a telescope from plywood, from plastic plumbing components, from ordinary items found in hardware stores. Even the most difficult component, the precision lens, or mirror, that gathers and focuses starlight into a precise image, should cost you no more than the very smallest commercially built telescope. If you prefer, you can grind and polish your own mirror or lens for the pleasure of doing everything important in building your telescope yourself.

What will a telescope do for you? There's no simple answer; what you get from a telescope depends largely on what you put into using it. However, any of the telescopes described in this book will enable you to discover for yourself the cratered surface of the moon and the appearances of the planets. You can come to know Jupiter's cloud belts and Red Spot, the mini-solar system of the four Jovian satellites, the rings of Saturn, the phases of Venus and Mercury, the polar caps and some of the surface markings of Mars. You'll be able to see sunspots, spot asteroids, watch bright comets when they appear. You'll be able to find Uranus and Neptune, although only with a very powerful telescope can you seek out tiny, distant Pluto.

You'll be able to see the colors of stars, stars that are double and stars that vary in brightness. You can search for the young, open clusters of stars nestled among the star clouds of the Milky Way and the ancient globular clusters that orbit the galaxy. You'll be able to see nebulae where stars are being born and gassy shells thrown off by dying stars. The views won't be in vivid color—the nebulae are too dim to excite the color-sensing cells in your eyes—but they're live views direct from the universe.

Looking deep into space, you'll be able to see galaxies as faint smudges of light against the glow of the night sky, and if you look carefully, knowing where to look and what to look for, you'll perhaps begin to see their spiraling arms. Even a few distant clusters of galaxies and a quasar are within the range of the telescopes in this book.

What must you give to gain these pleasures? First, it'll be up to you to learn the stars and constellations so you can find your way around the sky. You'll need to learn enough basic astronomy that you begin to understand what it is that you're looking at—otherwise the views are sure to disappoint you. Depending on your age and circumstances, you'll need understanding parents, an understanding spouse, an understanding landlord, or understanding next-door neighbors. You'll need a few hundred dollars to spend on parts and supplies for your telescope. You'll need patience while you wait for parts to arrive. You'll need workshop

Beyond a certain point, more magnification produces only a smaller field of view and a progressively fuzzier image.

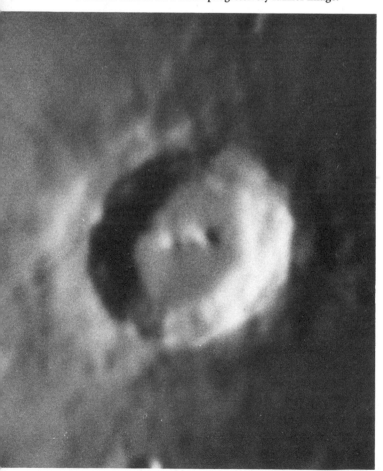

space while you're building it. When it's finished, you'll have to get up at strange hours to see Mercury and Venus rising before the sun and you'll have to stay up late to see Jupiter, Saturn, and Mars. You'll have to be willing, if you live under the glow of city lights, to treat yourself to weekends of camping far away, where the sky is truly magnificent.

You need to understand that a telescope won't do anything for you by itself. It's only what you see through it, what falls on your retinas and is interpreted by your mind, that matters. Your telescope will not be a magic carpet sweeping you away, but a tool for your mind and imagination, a tool you can apply for the exciting exploration of space.

MAGNIFICATION

All telescopes, regardless of construction or type, have two basic purposes: first, to provide magnification and, second, to provide light-gathering power. Magnification is the increase in the angular size of objects you view through the telescope. The moon, for example, appears small in the sky to the naked eye. Viewed through a telescope at a magnification of 50x, it would cover an angle about the same size as a medium-sized pizza held at arms length—more than sufficient to reveal the seas and craters on the moon's battered face.

The magnification of a telescope isn't fixed. Simply be removing one eyepiece, the small lens you actually look into, and putting in another, you can get pretty nearly any magnification you want with any telescope. For example, you can sweep along the Milky Way finding faint clusters of stars at a magnification of 24x, then switch eyepieces and split the components of a close double star at 300x.

What magnification is best? The answer is, only as much as required to do what you want to do and no more. Magnification is a two-edged sword: The more you magnify, the smaller the region of sky you see. With much over 100x magnification, you see only part of the moon when you look at it with a telescope. At 500x, it becomes difficult to bring celestial objects into the diminishingly tiny field of view.

Furthermore, the image becomes dim and hard to see at high magnification. Tiny residual shakes and

"Light-gathering" means that telescopes can reveal details of faint
celestial objects such as the core of the Orion Nebula, shown here.

shivers in the telescope's mounting are magnified also;
you may find, to your dismay, that the star you want
to examine is bouncing wildly around the field. When
you use binoculars, you realize that a wide, steady
field of view at a low magnification is better than a
high-powered but always-jiggling view.

The brighter planets show off their basic characteristics with 60x. Jupiter's cloud bands and Great
Red Spot require a little more, but the phases of Venus
need less. You can see Saturn's rings and the polar
caps of Mars quite well with 100x, although for careful
study, you may use upward of 250x. For nebulae and
star clusters, oftentimes the *lower* the magnification,
the better. You'll want the wide, bright view that only
the lowest magnification can give. At 60x, for example,
you can't view the whole Pleiades star cluster at once.
You'll seldom want more than medium magnifications for any deep-sky observation.

LIGHT GATHERING

But magnification is only half of the story. Celestial
objects are faint, hard to see, or even so dim they're
entirely invisible. A telescope serves to gather more
light than your eyes alone can. The telescopic image
will be not only bigger, but it will reveal fainter
celestial objects.

Light-gathering power is the ratio between the area
of the main light-gathering mirror or lens of the
telescope and the area of the pupil of the human eye.
The pupil of a young person opens to about 7 millimeters; by age 50, this will be only 5 millimeters. It's
best to take an average figure—about 6 millimeters—as
the pupil diameter of the "average" observer. A 60mm
lens or mirror, with a diameter 10 times and therefore
an area 100 times that of the "average" eye, has a
light-gathering power of 100.

Lunar craters Theophilus, Catharina, and Cyrillus seen with low-resolving power (bottom) and high-resolving power (left). Left photo by Jean Dragesco.

Compared with magnifications, which are lower than people usually expect of a telescope, light-gathering powers are impressively large. Even an ordinary 7x binocular with 35mm or 50mm lenses gathers dozens of times more light than your eyes can, revealing proportionately fainter stars. Telescopes quite feasible for a beginner to build gather a thousand times more light than your eyes can.

However, this does not mean that everything you see through your telescope will be brilliant. Light gathering is necessary to offset the dimming effect of magnification. As it magnifies, the telescope "spreads" the light from a celestial object, such as a nebula, over a larger angle, making the object appear not only larger, but also dimmer. Only at the very lowest magnification, its "normal" magnification, can a telescope show a scene with its normal brightness, as bright as it would appear to your unaided eyes.

If a telescope doesn't make things brighter, how then can it aid you in seeing faint celestial objects? A faint patch of light in the sky, such as a galaxy, is difficult to see if it is small; when it appears larger, as it does through a telescope, your eyes detect the enlarged patch more readily. The net effect is that you can see fainter objects with a telescope than you can without one.

Only stars, which are too small and too far away to be seen as anything but tiny disks of light in any telescope, grow in brilliance. Up to a point, magnification does not make the images of stars larger, so stars become brighter with the light-gathering power of the telescope. With a telescope having a light-gathering power of 100, you can see stars 100 times fainter than those you can see with the naked eye.

What is the faintest star you can see with a telescope? Astronomers rate stars on a magnitude scale. Typical bright stars in the sky are first-magnitude stars; the faintest visible to the naked eye are sixth. For each factor of 2.5 in light-gathering power, you can see stars that are one magnitude fainter. A 6″ telescope, such as those described in Chapter 5, 6, and 8, with a light-gathering power of 625 reaches stars of magnitude 13.5, seven magnitudes fainter than you can see with the naked eye. There are roughly 10,000,000 stars in the sky brighter than that limit.

APERTURE AND RESOLVING POWER

The two principal properties of a telescope, its magnification and its light-gathering power, depend on the diameter of the main mirror or lens, or its *aperture*. The bigger the aperture, the more powerful the telescope, both in its capacity for providing magnification and in its light-gathering power.

Another property of the telescope that depends on its aperture is its resolving power. The larger the aperture, the better a telescope is able to distinguish, or "resolve" as separate, close sources of light. Because of the wave nature of light, no telescope, no matter how perfect its optics, produces a pinpoint image of a star; instead, it produces a tiny spot, or "spurious disk." This is not the real disk of the star. The real disk is much too small for *any* terrestrial telescope to resolve; the disk you see is an effect caused by the telescope due to the diffraction of light, and it limits how much detail you can see with it. Suppose you're looking at two stars that are closer together than the spurious disk. You will see a single blur of light and won't be able to resolve, or separate, the two stars. The resolving power is the angular distance between the closest two equally bright stars that a skilled observer can readily distinguish as two stars rather than one star "unresolved."

If you think of any light source, such as the moon, as being made of many individual diffraction disks blended together by the telescope, it's clear that the better the resolution, or the greater the "resolving power" of the telescope, the finer the details of the lunar surface that will be seen. In other words, the bigger the aperture of the telescope, the more detail you can see with it.

Diffraction is a fundamental limit: There is absolutely no way to beat it. Aperture is, therefore, the most important single factor in any telescope's performance. When astronomers talk about telescopes, the first thing they mention is the aperture. "Yup, it's a six-inch Newtonian"—or refractor, or Schmidt-Cassegrain. Aperture comes first, whatever the type.

You're approaching the maximum useful magnification when the telescopic diffraction disk is large enough for you to see clearly. An observer with sharp

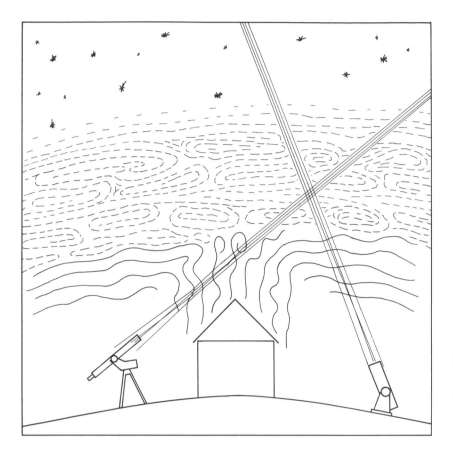

Before it reaches your telescope, light from celestial objects must pass through miles of Earth's turbulent atmosphere.

vision can distinguish two points about $\frac{1}{30}$ of a degree apart. Suppose you were observing with a 6″ telescope, which forms an Airy disk of $\frac{1}{3840}$ of a degree across: A magnification of 128x would enlarge the Airy disk enough for you to distinguish it as a disk. Fine detail is, of course, easier to see when it's made double, or even triple, the lowest magnification necessary. You'll find, however, that the sharpest, crispest, and most satisfying observations are those made at relatively low magnification.

Diffraction is only one of the reasons magnification cannot be increased without limit. Atmospheric turbulence and air turbulence inside the telescope tube enlarge stellar images beyond the size of the Airy disk. At the very best sites in the world—atop high mountains—images are occasionally smaller than 0.3 seconds of arc. In most places, however, the "seeing disk" is 1.0 second of arc or larger. This means that under ordinary conditions, a telescope with an aperture of 6″ to 10″ and a magnification of only several hundred

power reveals all or most of the fine detail that *can* be seen with any telescope.

Bigger telescopes are useful primarily because they gather more light, thereby reaching fainter stars and giving a brighter image of faint celestial objects. If you observe long enough with a big telescope, sooner or later there will be a wonderful moment when the turbulent atmosphere settles down for a few seconds and you'll glimpse more detail than you ever could with a small telescope.

TELESCOPE TYPES

Reflecting telescopes employ a curved mirror to reflect starlight together into an image. While there are many subtypes of reflecting telescopes, the most common is the Newtonian variety, invented by and named after the great English physicist, Sir Isaac Newton. Sir Isaac constructed his first Newtonian

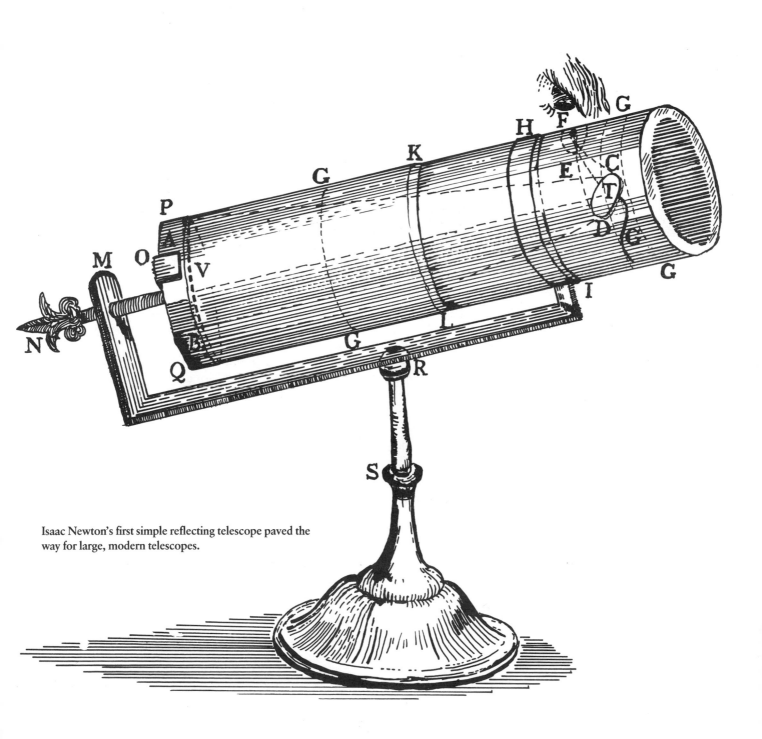

Isaac Newton's first simple reflecting telescope paved the way for large, modern telescopes.

telescope from a shiny metal mirror of 1⅓″ diameter that he ground and polished himself. He placed the mirror at the bottom end of a small cardboard tube. Light entered the tube, struck the curved mirror, and bounced back toward the front of the tube. Because of the curve in the mirror, the rays of light converged, or came together toward a focus. A small, flat mirror, called the diagonal, near the front of the tube bounced the light to the side of the tube, through a small viewing lens, or eyepiece, and into the observer's eye. His telescope magnified about 30x.

Modern Newtonian telescopes follow a similar plan. The major difference is that today's mirrors are not made from metal, but from glass, which is relatively easy to grind and polish. The glass is vacuum-coated with a thin layer of shiny aluminum metal. We also use precision multi-lens eyepieces for viewing the image rather than a single lens as Newton did. All the telescopes described in this book are much larger and more powerful than Newton's first reflector.

Newtonian reflectors have long been the choice for a first telescope. They are generally the simplest of all telescopes to build. There is only one major optical element, the primary mirror, and it is easy to make. In addition, today telescope mirrors are inexpensive relative to earning power: In 1947, a 6″ telescope mirror cost $60, but over three decades later—despite inflation—a mirror costs $80. In the past, nearly all would-be telescope-builders ground and polished the main mirror of the telescope, an act as much a matter of economy as of pride.

Another type of reflector, invented by Guillaume Cassegrain, uses a large concave mirror paired with a small convex secondary mirror. The image is formed behind the primary mirror; it reaches focus by going through a hole cut in the primary mirror. This type offers the astronomer a shorter tube length than that of a Newtonian at the cost of considerably greater difficulty in construction. Almost all large professional telescopes are of the Cassegrainian type.

Refracting telescopes use a lens to bend, or refract, starlight together to a focus. Early refracting telescopes, those made throughout the 1600s and into the 1700s, used a single glass lens. Because all colors of light are not bent identically by a lens, a single lens forms an image with a severe defect called "chromatic aberration." In practical terms, stars have big blue halos

unless the bending of the light through the lens is very slight. So astronomers made the lenses very weak. The resulting telescopes were very long—often 20 to 30 feet—and the longest were 150 feet. For all the apparent clumsiness of these telescopes, however, seventeenth-century astronomers made many important discoveries with them.

By the middle 1700s, lensmakers discovered that by combining two lenses made of different types of glass, about 95 percent of the chromatic aberration of a single lens could be canceled out. A two-element lens is called an "achromat," or a lens without color. However, achromats are hampered by the residual five-percent "secondary spectrum"—but again, the greater the distance between the lens and the image it forms, the less significant is chromatic aberration. This means that refractors are still relatively lengthy telescopes, typically 15 times the diameter of the achromatic lens.

The lenses are mounted together in a cell, or holder, at the upper end of the tube. Starlight enters the lenses, and after being bent by each of them, slowly converges near the bottom of the tube to form an image that the observer views with an eyepiece.

Refractors have been the classic starting telescopes for many generations of astronomers. Refractors with small apertures, while not offering the light-gathering power of a large telescope, are highly portable, light-weight, and require little or no maintenance.

Refractors offer a number of appealing features. The aperture is not obstructed by a Newtonian diagonal or a Cassegrainian secondary mirror, so the image quality is superior. Since the tube of the refractor is closed at both ends, air currents inside the tube itself do not disturb the quality of the image. Given care, a refractor requires little maintenance, while reflectors often need periodic cleaning and realignment of their mirrors. Despite these pluses, large-aperture refractors have one big minus: They are expensive. Reflectors offer much more light-gathering power and higher theoretical resolution for the same amount of money.

Catadioptric (lens-plus-mirror) telescopes have become popular since World War II. The two kinds of catadioptrics, the Maksutov Cassegrainian and Schmidt-Cassegrainian, are variants of the Cassegrainian telescope. While the two mirrors do most of the work of bringing the light to a focus, the lens at the front of

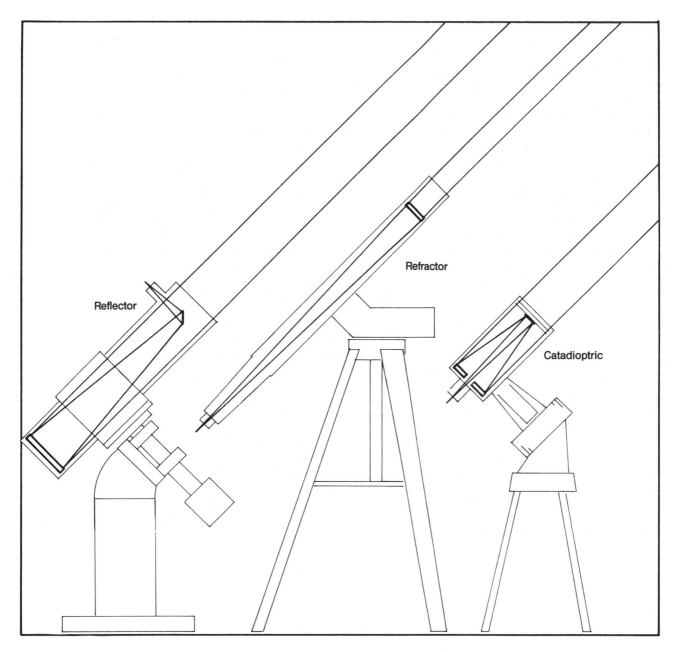

Reflectors, refractors, and catadioptric telescopes are the major types in use today.

the telescope corrects small residual image errors. The Maksutov type uses a thick, steeply curved lens of high-quality optical glass to produce its corrections, while the Schmidt type uses a thin, nearly flat corrector with a complex shape polished into its surface.

Catadioptric telescopes offer good optical performance in a very short optical tube. Several commercial firms manufacture these types of telescopes; they are viable largely because the makers mass-produce the tricky curves of the correctors. Nevertheless, catadioptrics are much more expensive than Newtonians of the same aperture. Neither type of catadioptric is a first telescope project.

TELESCOPE MOUNTS

Even the best telescope has no value unless you can point it quickly and easily toward celestial objects you'd like to see and trust it not to shake or vibrate once it's pointed. The same magnification you apply to the image in a telescope is also applied to every shake, wiggle, and shimmy in the structure that supports and directs the telescope. The mounting of your telescope, first and foremost, must be vibration-free and solid if you expect to realize the instrument's potential.

If the primary feature of a successful telescope mounting is stability, what comes after that? First, especially for a new astronomer, the mounting should be easy to use. Learning where to look for stars, nebulae, and galaxies is effort enough, but to cope additionally with a complicated telescope mounting in the dark leads to frustration. The more time you spend making adjustments and alignments, or fiddling with wires and knobs, the less time you'll spend looking at the sky.

Second, unless you're going to build a permanent observatory in your backyard, both mounting and telescope should be portable. Why? Each time you observe, you'll have to carry them outside and set them up. If your observing area has a lot of trees, you may have to move the telescope in order to see things in different parts of the sky. Portability means that the mounting should be lightweight and reasonably compact, easy for one person to carry and set up alone.

Next, the mounting should allow the telescope to move smoothly. As you follow a star or search out a faint, distant galaxy, you'd like the mount to respond to your actions. You shouldn't need to release a clamp device, tap the telescope until it moves somewhat too far, tap it back, then clamp it in place again, only to have it jiggle away from the object you're trying to observe as you turn the clamp. If you've ever used an inexpensive import telescope, you know the routine.

Finally, if possible, it's nice if the telescope mounting will follow the stars for you. But that's a luxury, at least until you've spent quite a bit of time in observing; and you can do without automatic tracking, especially if having tracking means sacrificing stability, portability, or ease of use.

No single telescope mounting can provide all the desired features fully, just as no one telescope can handle all observing tasks with equal competence. The most stable mountings don't track automatically, but they are light and exceptionally easy to use. Mounts with motors that allow them to follow the motion of the sky (called equatorial mounts) tend to be heavy, nonportable, and tricky to set up and operate. However, who ever said that a telescope should be used on just one mounting? If you plan things right, you can use one telescope on several different mountings.

I recommend a light, portable, solid mounting when you first build your telescope, to get you started on celestial observing right away. As your interest matures, such a mounting will still be ideal for quick and casual observing from the backyard or up in the mountains far from civilization and city lights, even if later you've replaced it for "serious" backyard use with an equatorial mount. Still later you might decide to machine a massive and stable mount equipped with a precise drive for hour-long, deep-sky photography, a project likely to take you several years.

In Chapter 3, we'll go into detail on the subject of mountings, but for the present, this should be sufficient to start you thinking ahead.

PICKING THE RIGHT TELESCOPE FOR YOU

Before deciding which kind of telescope to build, consider the three primary criteria for choosing a telescope: its aperture, its physical size, and its cost. If choosing *only* on the basis of aperture, which determines the magnification, the light-gathering power, and the resolution, bigger is better; but if you consider size, cost, and aperture together, you'll probably reach a rather different conclusion.

Think about how you live and how the telescope will, quite literally, fit into your life. How big *can* it be? Where will you observe? How will you get your telescope there? If you must drive, how wide is the backseat of your car? Do you own a station wagon? Can you carry the telescope up the back stairs? Into the elevator? If it's eight feet long, you'll need help. Does your yard have trees? How far will you need to carry a telescope in order to use it? Where will you store it? Will it be easy to take outside?

If your telescope is not convenient to use, you won't use it. Face that fact, plan accordingly, and don't get your heart set on a telescope that's not going to be practical for you.

The 4″ reflector, for example, is excellent for beginners—either kids or adults—who want a low-cost telescope. The 4″ reflector described in Chapter 4 makes a fine parent/child project for kids from 8 to 12. Those 12 and older can do much of the work themselves. With a 4″ telescope, you can see all the major sights in the solar system—the belts of Jupiter, the rings of Saturn, the polar caps of Mars. You'll be able to split lots of double stars, to see variables, star clusters, and even a sampling of galaxies. Because the light-gathering power is relatively limited, however, faint deep-sky objects are difficult or impossible to see. The total cost is around $100.

The 6″ Newtonian reflector aperture is often the telescope recommended for someone building a first telescope. It's perfect for a high-school student or an adult who wants a relatively portable telescope. An observer with a 6″ reflector can see features on Mars, knots and swirls in the cloud belts on Jupiter, and the narrow Cassini Division in Saturn's rings. With diligent searching, even from a city/suburban location, you can find all of the nebulae and star clusters included in the list of deep-sky objects compiled by Charles Messier in the late 1770s, which is still the standard fare of modern observers. A good 6″ telescope can keep you busily observing for many years. Figure on spending about $250 if you buy the mirror finished

and do most of the other work yourself.

The 10″ telescope is a jump into another class. Solar-system objects can be seen in considerable detail; globular clusters appear as balls of stars rather than fuzzy glows; the number of galaxies within reach expands dramatically; and the greater light-gathering power brings out the spiral structure of the brightest. A 10″ telescope can keep a serious observer satisfied, but the telescope will still be portable enough to carry out and set up each time it's used. A 10″ telescope will cost about $400, but an equatorial mounting suitable for astrophotography could total well over $1000, and if you start thinking about an observatory for it, figure on another $500 or more.

The 6″ f/15 refractor described in Chapter 8 is a powerful instrument, a telescope for a lifetime. If you build one though, it will be over eight feet long, bulky, and quite heavy. It can provide images equal to those of a somewhat bigger reflector; but, of course, its light-gathering power will still be that of a 6″ aperture.

What about still bigger—really big—telescopes? A 10″ begins to exhaust the magnification that atmospheric seeing normally allows, so the gain from a larger aperture is mainly in light-gathering power. After a 10″, the next big step is to a 16″, for which the number of galaxies available jumps from the hundreds into the thousands. But don't underrate a 10″ telescope. There is an enormous amount to see with one. If you must have a big telescope, start with a 10″ and use it for a year or two. Learn the sky first, then make the big one your second telescope.

The main thing now is to start simple. If you try to build the ultimate rig from scratch, you could get bogged down, never finish it, and never even get a glimpse of the sky for all your efforts.

How Telescopes Work

2

Before you can understand how a telescope works, you must know some basic telescope terminology. The principal optical component of any telescope is its *objective*. The objective may be either a lens or a mirror. If it's a lens, you may hear it called the objective lens, object glass, or simply "glass," although sometimes the term "glass" also refers to an entire refracting telescope. If the objective is a mirror, as it is in a Newtonian reflector telescope, it may be called the *primary mirror*, the primary, the speculum (which is merely Latin for "mirror"), or the objective mirror.

In a Newtonian, the small mirror that reflects the converging light out the side of the telescope tube is referred to as the *diagonal mirror*, the diagonal, or the Newtonian secondary. In Cassegrainian-type telescopes, a small *secondary mirror* reflects light back toward the primary and through a hole in it. The deeply curved lens in a Maksutov-Cassegrainian is the *correcting lens*, but in a Schmidt-Cassegrainian, it's sometimes called the *correcting plate* because of its thinness.

The *aperture* of a telescope is the diameter of the beam of light that enters the telescope and eventually reaches the eye. It is normally equal to the diameter of a telescope objective, although in other optical systems, such as camera lenses, the front element of the lens may be considerably larger than the image-forming beam of light.

Light that enters a telescope comes to a *focus*, a place where all of a star's light that has fallen on the objective adds together. If you were to place a sheet of paper or a piece of finely ground glass at the focus, you would see the light from the star as a tiny, bright speck. Light from stars or from an object such as the moon forms an image that can be viewed on a sheet of paper or on ground glass. If you place photographic film at the focus of a lens or mirror, the chemical action of the light on the film will create an image in the film.

The distance between a simple lens or mirror and the focus is the *focal length*, or focal distance, of the mirror or lens. The depth of the curved surfaces of a mirror determine its focal length. Deep and strongly curved mirrors have a short focal length, while those that have weak, or shallow, curves have long focal lengths. The same is true of lenses, except that the

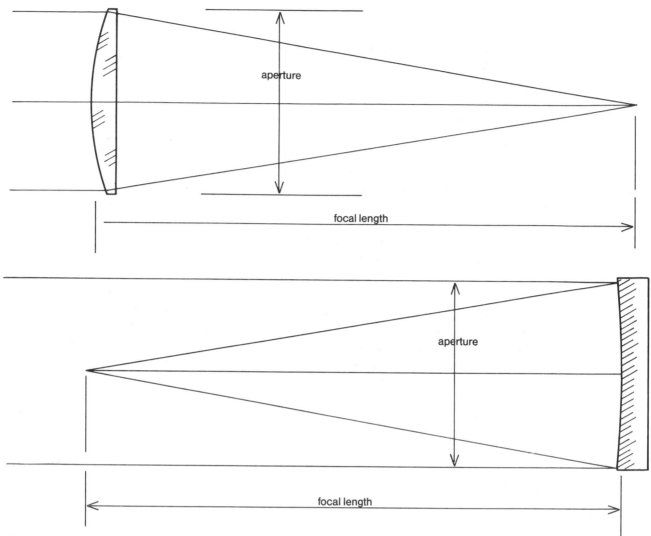

For both lenses and mirrors, the focal ratio is the focal
length divided by the aperture.

type of glass they're made from also affects their
focal length.

A compact notation for describing the focal length
of a telescope is the ratio between the focal length
and the aperture, the *f/ratio* or *f/number* (read as EFF-
number). A telescope with an aperture of 6″ and a
focal length of 48″ has a focal ratio of 8 and is an f/8
mirror. The f/number is sometimes called the *relative
aperture*, since it tells us the aperture *relative to* the
focal length.

An astronomer uses a combination of small lenses
to examine the image formed by the objective of the
telescope. These lenses, held in a suitable metal barrel,

compose the *eyepiece* (American usage), or *ocular*
(British usage). Eyepieces are interchangeable—you
simply slide one out of the eyepiece holder and slide
in another. Each eyepiece, like the objective, has a
focal length, usually between a fraction of an inch and
several inches. Ordinarily the focal length is con-
spicuously engraved on the top of the eyepiece.

The *magnification* of a telescope is simply the ratio
of the focal length of the telescope to the focal length
of the eyepiece. By changing to an eyepiece with a
different focal length, you change the magnification.
Thus a telescope with a focal length of 48″ used with
an eyepiece of 1″ focal length gives a magnification of

48. If you put in another eyepiece, one with a focal length of ½″, the magnification will become 48″ divided by ½″, or 96x. Eyepiece focal lengths are almost always given in millimeters rather than in inches, and before carrying out the division, you should make sure you have both focal lengths in the same units. Forty-eight inches are 1220 millimeters; thus, used with a 6″ f/8 telescope, an 8-millimeter-focus eyepiece produces a magnification of 1220 divided by 8, or 153x.

The area of sky visible through an eyepiece is the *field of view*. The angle of sky covered is the *real field of view*; the angle you see when you look into the eyepiece is the *apparent field of view*. If the field of view of your telescope just barely takes in the entire full moon, which is ½° across, the real field of view will be ½°. However, when you look into the eyepiece, the field appears much larger to you. To a first approximation, the magnification times the real field of view equals the apparent field of view. Thus, if you were using 80x magnification, your telescope would show a ½° field approximately 40° in apparent diameter.

Light enters the aperture of the telescope, but where does it exit? Just behind the eyepiece there is an image of the objective formed by the eyepiece. All the light in the field of view of the telescope leaves through it. This image is the *Ramsden disk*, or the *exit pupil*. When you look through the telescope, what you're doing is placing your eye in the exit pupil to receive light from the telescope. The diameter of the exit pupil is the aperture of the telescope divided by the magnification. Eyepieces that produce different magnifications must, therefore, produce different exit-pupil sizes.

The pupil of your dark-adapted eye probably opens between 5 and 7 millimeters. If the exit pupil of the telescope is larger than the pupil of your eye, your eye won't admit all the light from the telescope. This means that the lowest useful magnification for any telescope is the magnification that produces an exit pupil of 6 millimeters. This is approximately 4x per inch of aperture, or about 24x for a 6″-aperture telescope.

When the exit pupil is just the size of the pupil of the eye, the telescope gives a view of the same brightness as the view without a telescope, although, of course, nebulae look larger and are therefore easier to see with a telescope. This rock-bottom magnification is the "*normal*" *magnification*. Any telescope operating at the normal magnification is a *richest-field telescope*, or RFT, because at low magnification and with a wide field of view, the field is rich with stars and nebulae.

A simple telescope: The ratio of the focal length of the objective and the focal length of the eyepiece is the magnification.

$$\text{Magnification} = \frac{\text{focal length (objective)}}{\text{focal length (eyepiece)}}$$

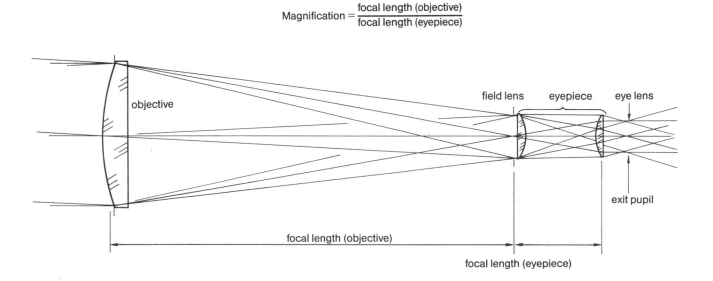

objective

field lens   eyepiece   eye lens

exit pupil

focal length (objective)

focal length (eyepiece)

PHOTONS, WAVES, AND RAYS: THE NATURE OF LIGHT

Telescopes are devices that control light, but what is light? Without pretending to define light in any fundamental way, we can imagine it both as packets of energy and as a rapidly oscillating electrical and magnetic wave. Light sometimes behaves like particles interacting individually with atoms; at other times, oscillatory effects predominate and light behaves like a wave. The wave and particle behaviors are not contradictory, but are manifestations of the same rather complex phenomenon.

Each packet of light, or photon, has a discreet energy, wavelength, and frequency associated with it. These vary over an enormous range. The most energetic photons, with energies of tens of millions of electron volts and wavelengths of 10^{-13} meter, are gamma rays and X rays; the least energetic are heat radiation and radio waves with wavelengths extending many kilometers. When the photons have an energy of 2.5 electron volts, or 4 trillionths of an erg per photon, the wavelength is 500 billionths of a meter and the frequency is 500 trillion cycles per second. These are photons of yellow light. Given some 30 percent more energy, it's deep violet light; given 30 percent less, it's deep red. Outside that range, photons are not visible.

Waves of electromagnetic energy spreading from an oscillating electron are sometimes compared to waves of water spreading from a pebble thrown into a pond. Yet there is no counterpart of photons in the pebble-in-the-pond analogy. Instead, we speak of rays emanating from a star, traveling through space, entering a telescope, reflecting off its mirrors, traveling through the ocular, and entering our eyes. Light rays follow straight paths until they bounce off a mirror or pass into a transparent material such as glass. If you think of the wavefront as the crest of the wave spreading outward from a pebble tossed into a pond, "water rays" would be straight lines extending from the spot where the pebble went in. Rays are an artificial device but very useful in speaking about light.

Because photons are wavelike, they don't sum in a simple fashion—they add "in phase." You can think of them as having crests and troughs, like waves in water. If they add in phase, crest to crest, they add constructively. If they meet crest to trough, they cancel each other out, or add destructively. If the total distance every photon of light from a distant star travels inside the telescope is exactly the same, the photons will meet at the focus and add constructively. If there are deviations of a fraction of a wavelength of light in the total path, then some photons will take a longer path and arrive at the focus out of phase, hence add destructively.

This explains why telescopes must be optically precise: If the mirrors or lenses have errors bigger than a fraction of a wavelength of light, the photons do not add in phase.

But it's not so simple. Even if a telescope is optically

Radio, X rays, and light are forms of electromagnetic radiation.

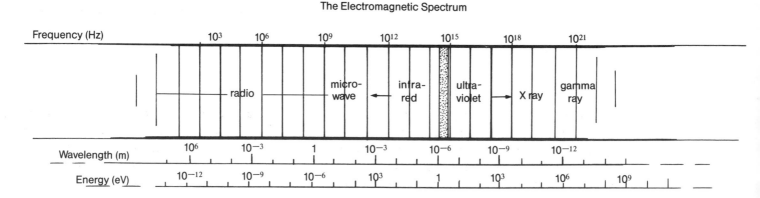

The Electromagnetic Spectrum

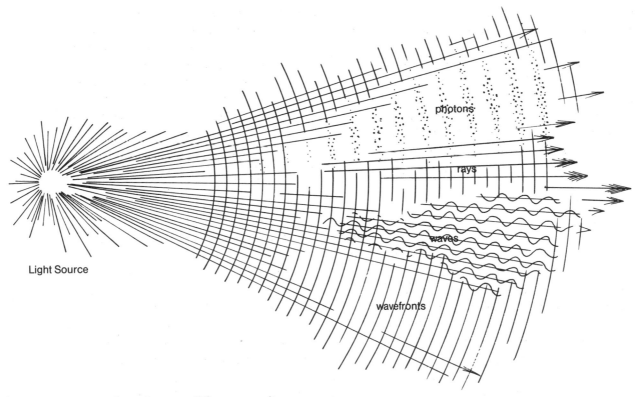

Light Source

Photons, rays, waves, and wavefronts are different ways of conceptualizing one phenomenon—light.

perfect, photons arriving at any small section of the aperture spread as if they'd originated there. What happens when the photons from one side of the aperture meet photons from the other side? In a perfect telescope, the pathlengths are equal so they meet at the focus in phase.

Next, let's look immediately beside the focal point. At this new location, because we are farther from one side of the aperture and nearer to the other side, the waves arrive a little out of phase. Their intensity is diminished but still greater than zero.

This means that light will not form a perfect geometric focal point; some energy appears around it. As we move away from the focal point, the path difference between the photons from the opposite sides of the aperture grows and the intensity of the light drops. When the path difference is ½-wave, the photons sum out of phase, so the intensity falls to zero.

As we move farther, the path difference reaches one wavelength and photons again add constructively.

The intensity at this point is, however, only about 2 percent as intense as the geometric focus because if we sum the phases from all the sections of the aperture (and not just from opposite sides), the sum is small.

The mathematical description of the distribution of light in our ideal telescope is the *point spread function*; the physical reality is an intense central region called the *Airy disk* (named after Sir George Airy, who first derived the equation for the point spread function), which is encircled by *diffraction rings*.

Bear in mind that the Airy disk is quite small. Its diameter, to the first dark ring, expressed in radians, is:

$$\text{Airy disk}_{rad} = 1.22 \ \lambda/D$$

or in seconds of arc:

$$\text{Airy disk}_{arcsec} = 251{,}600 \ \lambda/D$$

where D is the diameter of the objective and lambda,

λ, the wavelength of light given in the same units. For a telescope with an aperture of 150mm (= 6″) and yellow light (lambda = 0.00055mm), the Airy disk is 0.0000044 radians, or 0.923 seconds of arc, across. The linear size of the Airy disk equals the angular diameter times the focal length of the telescope. For a telescope with a 48″ focal length (1220mm), the Airy disk works out to be 0.0053mm in diameter—or about 0.0002″. No wonder star images usually seem to be tiny pinpoints.

REFLECTION AND REFRACTION

Light interacts with matter in many ways. Perhaps the simplest to understand is *absorption*. A piece of black cloth does not reflect most of the photons that strike it. Instead, the atoms in the cloth will then gain the energy of the photon, and the cloth becomes warmer as a result. Instead of absorbing light, some materials reflect it; i.e., photons bounce off these materials. Surfaces that reflect most of the photons striking them look white. Absorption and diffuse reflections are useful as ways of getting rid of unwanted light inside a telescope.

If a light-reflecting surface is smooth and metallic, or mirrorlike, the reflection is *specular*. Unlike photons after a diffuse reflection, photons reflected from a mirror surface retain information about their original direction and bounce off of the surface according to an exact geometric law:

$$\theta_{reflected} = \theta_{incident}$$

We measure these angles from the *normal*, a line that is perpendicular to the surface. On curved surfaces, the normal is perpendicular to lines that are tangent to the surface.

Because the behavior of photons reflected from them is predictable, mirrors are useful for controlling light in a telescope. Suppose that we make a concave ("-cave" = inward sloping) mirror. We know that light striking it will converge toward the focus. A flat mirror will redirect light rays without otherwise altering them.

Light behaves in a different way when it passes into a material such as glass. As it enters the new, denser medium, the photons slow down, and the ray path bends too, or is *refracted*. Refraction, like reflection, obeys exact geometric rules. The angle of the bending

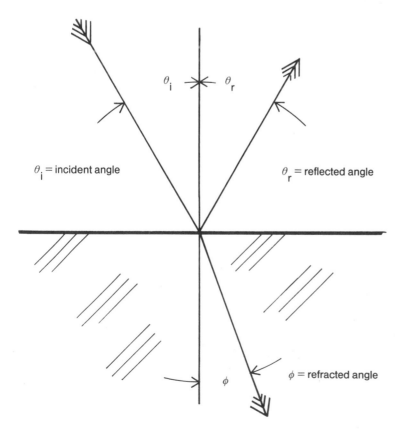

θ_i = incident angle

θ_r = reflected angle

ϕ = refracted angle

Light obeys the laws of reflection and refraction when interacting with matter.

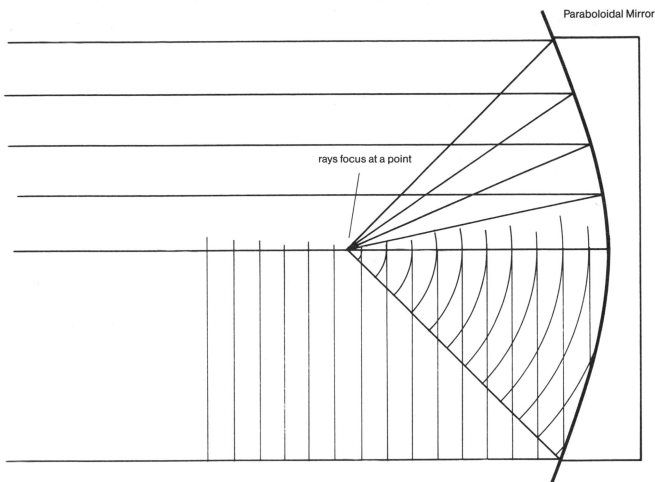

A paraboloid focuses plane wavefronts of light to a point. Ray paths are also shown.

bears a precise relation to the angle of the incoming light:

$$n_{\text{refracting medium}} \times \sin(\theta_{\text{refracted}})$$

$$= n_{\text{incident medium}} \times \sin(\theta_{\text{incident}})$$

where n is the *refractive index* of the medium through which the rays are traveling. The refractive index is the ratio of the velocity of light in a vacuum to the velocity of light in that medium. When light passes from air (n = 1.0003) into a denser medium such as glass (n = 1.52), the angle of refraction is smaller than the angle of incidence, which means that light is bent closer to the normal to the surface.

Refraction, like reflection, can be used to bend and focus light in a telescope. Light that passes through a biconvex piece of glass (shaped like a bean, or "lentil," hence our word "lens") will bend and at some point behind the lens come together, or converge, to a focus.

COMBINING PHOTONS

In order to bring photons together in phase, the distance from the focal point to any point on the incoming plane wavefront must be equal to the distance from any other point on the wavefront to the focus. The required curve must be one that lies equidistant between a plane and a point. Is there such a curve?

The answer is yes: In three-dimensional space, the curve is a *paraboloid* (puh-RAB-uh-LOID). A paraboloid is formed from all the points equidistant from a particular point and plane, just as a sphere is formed from the points lying at a constant distance from a particular point. Their definitions, in fact, imply that a sphere is quite similar to a paraboloid. This similarity is extremely valuable to opticians because a sphere is an easier shape to grind into glass or metal than a paraboloid.

How similar are these shapes? The equation of a parabolic section of a paraboloid with its axis on the x-axis is:

$$X_p = y^2/2r$$

where r is the radius vertex of the parabola (the distance from the generating point to the surface of the curve). The equation of a matching circle (a plane section of a sphere) with radius r is:

$$X_s = y^2/2r + y^4/8r^3 + 3y^6/48r^5 + \ldots$$

The first terms are identical, so the difference between a sphere and parabola is:

$$X_s - X_p = y^4/8r^3 + 3y^6/48r^5 + \ldots$$

The curves at the edge of a 6″ f/8 sphere and paraboloid lie about ½ wavelength of yellow light apart. Only a minor alteration to a spherical mirror can produce a paraboloid. The equations also tell us that the difference between the curves grows rapidly as y

(the radius of the mirror) gets larger. This implies that the larger the mirror, the greater the alteration necessary to modify a spherical surface into a paraboloidal one.

By now a question may be nagging at the back of your mind: What happens to photons when an optical system doesn't combine them in phase? They don't disappear; instead they turn up *near* the focus as scattered light. If the optical system is almost perfect, the airy disk may be only slightly enlarged; if the optical system is badly flawed, the images it forms will be fuzzy blobs.

TRACING RAYS

We have now seen that spherical and paraboloidal mirrors have the correct physical properties to focus a wavefront of light. It stands to reason that since light

The focal length of a sphere is half its radius of curvature, but a sphere does not focus light to a point.

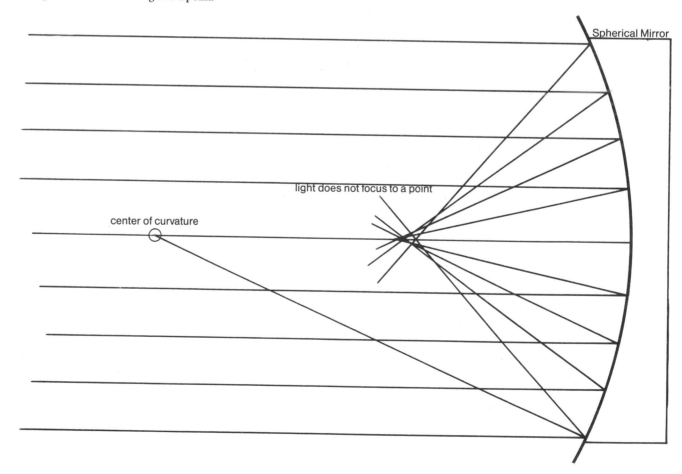

Spherical Mirror

light does not focus to a point

center of curvature

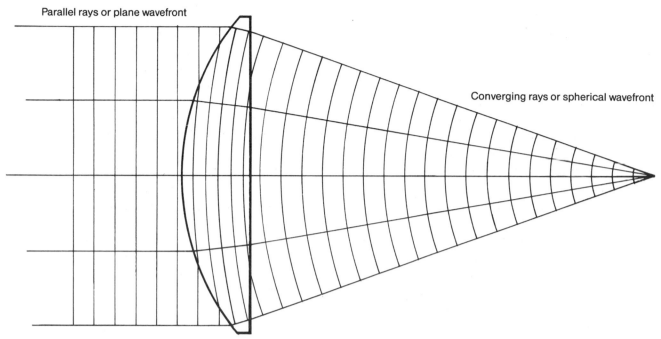

Parallel rays or plane wavefront

Converging rays or spherical wavefront

An idealized lens focuses light by bending light as it passes through it. Both rays and wavefronts converge to a point.

can be described not only as a wave, but also as a ray, we should get the same results if we trace the paths of light rays through a telescope. Let's now investigate the geometric properties of paraboloidal and spherical mirror surfaces that allow them to focus light rays to a focal point.

We'll begin with a simplifying assumption: For small mirrors and lenses, the difference between a sphere and a paraboloid is negligible. Strictly speaking, this is not true, but as we saw before, a spherical surface is a very good approximation of a paraboloidal surface.

Suppose we construct a spherical reflector so that its center coincides with a small source of light, such as a flashlight bulb. Since each ray follows a radius of the sphere, light from the source will strike the mirror at a 90° angle to its surface. According to the law of reflection, it will be reflected at a 90° angle also, and therefore all the rays will travel right back to the source. When we make a mirror, we use this optical configuration to test the quality of the mirror.

Let's remove the bulb and allow parallel rays from a star to fall squarely on the mirror. What will they do? According to the law of reflection, each ray will be reflected at the same angle from the normal as it was before it struck. The angle between each ray and the

optical axis is, to a very good approximation, twice the angle between the radius and the optical axis (for paraboloidal mirrors, this is an exact relationship). Each ray will thus intercept the optical axis halfway between the center of the sphere and its surface: This is the focal point. The focal length, f, of a mirror equals half its radius of curvature, r, or:

$$f = r/2$$

Thus a paraboloidal or spherical mirror satisfies the laws of geometric optics for forming an image, just as we have seen that it satisfies the condition that photons be combined in phase.

What does this tell us about the ideal shape of a wavefront as light comes to focus? Recall that the rays are perpendicular to the wavefront and that the light reaches focus in phase. Each ray is a radius extending from the focal point, and wavefronts are therefore spherical shells converging on it. This is exactly the inverse of the radiating point source of light we began with; the outgoing wavefront was a series of expanding spherical shells. In optics, everything can be reversed.

Suppose we grind and polish a thin disk of glass to

make a lens. On the front side is radius r_1, and on the back side, r_2. Both surfaces will alter the slope of each ray passing through it. The focal length, f, of the resulting lens is:

$$1/f = (n-1)\,(1/r_1 - 1/r_2)$$

This is the *lensmaker's formula*. It is important to remember in applying it that the formula is valid only if the sign of each radius is positive when the surface is convex to the light and negative when the surface is concave; for a lens that is convex on both sides, r_1 will be positive, but r_2, concave toward the light, will be negative.

This brief treatment only scratches the surface of geometric optics. Using pencil and paper, calculator, or home computer, it is possible to trace rays through complex optical systems with great precision. If you are interested in learning more, consult Appendix C for a listing of books on optics.

RESOLVING POWER AND DIFFRACTION

So far, we have looked at the behavior of light from a single point source as it interacts with mirrors and lenses. But what happens when there is more than one source? The photons focus at many points, each point having an intensity corresponding to a particular direction—an *image* of the scene before the lens or mirror.

Diffraction occurs because of the wave nature of light. Diagram not drawn to scale.

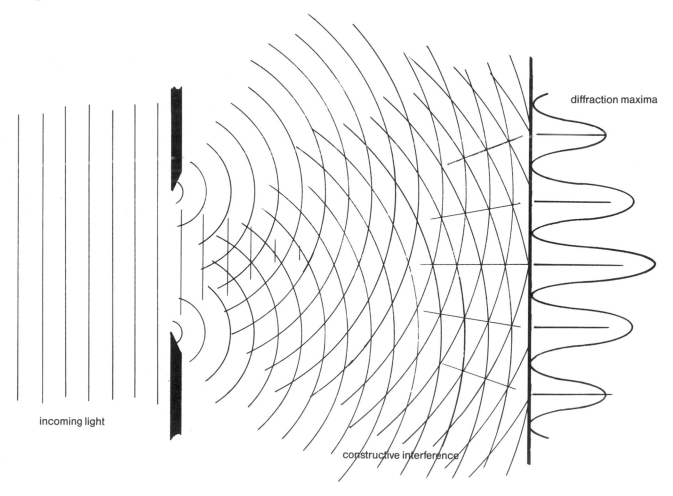

incoming light

diffraction maxima

constructive interference

Hold a magnifying glass in front of a sheet of white paper. Move it back and forth until you see a picture of the room form: the lamps, windows, furniture, people. The image is upside down, probably not very sharp, and dissolves into blurriness if you move the lens very much. All telescopes form images like this—a real picture that can be seen on a sheet of paper.

The location in the image depends on the angular direction to the light source. If two sources of light are separated by a small angle, which we will call θ, their separation at the focus is:

$$\text{separation} = f \times \theta \qquad (\theta \text{ in radians})$$

$$\text{separation} = f \times \theta/57.3 \qquad (\theta \text{ in degrees})$$

$$\text{separation} = f \times \theta/3438 \qquad (\theta \text{ in arcminutes})$$

$$\text{separation} = f \times \theta/206261 \qquad (\theta \text{ in arcseconds})$$

This is a useful formula for astronomy. Suppose we want to know how big an image of the moon our telescope will form. The moon has an angular diameter of ½° (or 30 arcminutes or 1,800 arcseconds). A 6″ f/8 mirror, having a focal length of 48″, will produce a lunar image 0.42 inches across. Although this may seem small, compare it to the diameter of the Airy disk (0.0002 inches!) and you will see that a great many Airy disks—about 2,100—and therefore a great deal of lunar surface detail, gets squeezed into that image.

Suppose we wish to observe two stars that are close together. When they are separated by several times the diameter of their Airy disks there is no difficulty whatever in distinguishing them. If we observe a closer pair, however, their diffraction rings overlap. If one star is considerably fainter than the other, its Airy disk might be lost in the light of the other's diffraction rings.

However, if the stars are more or less equal in brightness, we can readily distinguish them until the center of one Airy disk lies on the outer margin of the other's Airy disk. This separation we normally define as the *resolving power*, or *Rayleigh limit*. This limit, in seconds of arc, is:

$$\text{Rayleigh limit} = 5.5/D \qquad (D \text{ in inches})$$

$$\text{Rayleigh limit} = 138/D \qquad (D \text{ in millimeters})$$

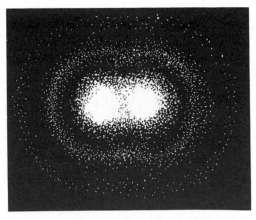

Two stars are "resolved" when their images can be recognized as two separate diffraction patterns.

where D is the aperture of the telescope. Note that this is exactly the same result that we obtained for the diameter of the Airy disk.

The resolving power of a 6″ telescope, for example, is most impressive. Substituting the equation above, we get 0.92 arcseconds—the angle between two touching blueberries a mile away! However, this definition of resolving power is arbitrary. Under good conditions, a skilled observer can distinguish two star images (and thus "resolve" them) when they are roughly 20 percent closer together than the Rayleigh limit. This is called the *Dawes limit* after the famous double-star observer, William R. Dawes, who established by experiment what he called the "separating power" of telescopes. The Dawes limit, in seconds of arc, is:

$$\text{Dawes limit} = 4.6/D \qquad (D \text{ in inches})$$

$$\text{Dawes limit} = 116/D \qquad (D \text{ in millimeters})$$

Fig. 2-12

Since very small optical pathlength errors, from poor telescope optics or a turbulent atmosphere above the telescope, can destroy the ideal shape of the Airy disk, a telescope and a night that will allow you to resolve stars down to the Dawes limit are a good telescope and a good night indeed.

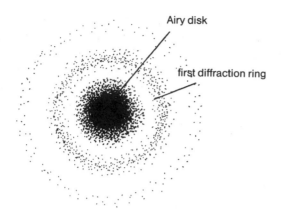

Airy disk

first diffraction ring

Even when formed by a perfect optical system, a star's image is spread by diffraction.

HOW NEWTONIAN REFLECTORS WORK

Newtonians are optically simple: Given an accurate paraboloidal mirror, a truly flat diagonal, and a well-designed eyepiece, designing a Newtonian telescope consists of little more than spacing these optical elements properly, relative to one another. However, there are several important details which you cannot ignore. How big should the diagonal mirror be in order to reflect all the light from the primary to the eyepiece? Where does it go?

The main rule in the optical layout of a Newtonian is that *the distance between the primary mirror and the eyepiece, measured along the path of the light rays, equals the primary mirror's focal length.*

$$F = L_{de} + L_{md}$$

where L_{de} is this distance from the diagonal to the eyepiece and L_{md} is the distance from the mirror to the diagonal. The diagonal mirror breaks the focal length into two parts which add up to the focal length of the mirror. Sounds great. But as soon as you try to apply this, you realize that the length and diameter of the tube that holds the optical parts also determines where the diagonal goes.

First consider the eyepiece to diagonal distance, L_{de}. This consists of three segments: the distance from the diagonal to the outside of the tube, the racked-in height of the focuser, and the amount the focuser is racked out:

$$L_{de} = T/2 + H + 1''$$

Two mirrors—one paraboloidal and one flat—that's all there is to a Newtonian telescope.

Here T is the outside diameter of the tube, H is the racked-in height of the focuser, and 1″ is a reasonable distance to allow for focus travel. Subtract this from the focal length of the mirror to obtain the diagonal-to-mirror distance:

$$L_{md} = F - L_{de}$$

Convert this figure to the more practical end-of-tube-to-diagonal length, L, by adding the end-of-tube-to-mirror-surface length, E, to it:

$$L = L_{md} + E$$

Finally, combine these equations to yield L directly:

$$L = F + E - T/2 - H - 1''$$

Diagonal Size. How large should the diagonal mirror be? The converging cone of light for a single star has diameter d at the diagonal. This means that the diagonal must be *at least* as large as d in its smallest dimension, which is called the minor axis:

$$d = (D/F) \times L_{de}$$

D is the diameter of the primary mirror and F its focal length. A diagonal this size will reflect all the star-light from one star to the focal plane, but not fully illuminate a wider field.

Suppose that you want a fully-illuminated field with angular diameter θ, measured in degrees. Its diameter d_f will be:

$$d_f = F \times \theta/57.3$$

For most purposes, a field of ½°—the same as the angular diameter of the moon—is quite adequate. To illuminate a field of diameter d_f, the following equation gives a good approximation of the diagonal size needed:

$$d = d_f + (D/F) \times L_{de}$$

For a 6″ f/8 Newtonian, then, given D = 6″, F = 48″, L_{de} = 8″, θ = 0.5°:

$$d = (48 \times 0.5/57.3) + (6 \times 8/48) = 1.419''$$

It would make sense to use the next size up, or a diagonal with a 1.5″ minor axis, just to be on the safe side. A slightly smaller diagonal, say one with a 1.25″

The optical layout of a Newtonian telescope.

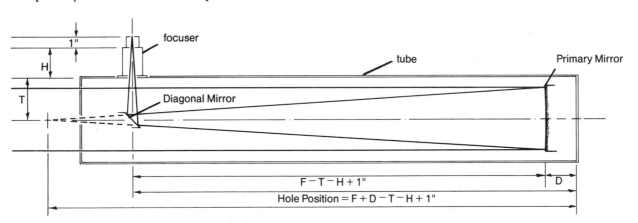

F − T − H + 1″

Hole Position = F + D − T − H + 1″

H = height of focuser

T = semidiameter of tube

D = distance between mirror and back of tube

or 1.30″ minor axis, would illuminate a slightly smaller field θ and you would not, in practice, notice the difference.

REFLECTOR ABERRATIONS

Until now we have pretended that mirrors and lenses are perfect. In the next two sections, we will look at the image-forming faults, or aberrations, that occur with mirrors and lenses. In the Newtonian, the major aberrations are spherical aberration, coma, and astigmatism; in the refractor, chromatic aberration is the principal difficulty.

Although spherical and paraboloidal mirrors are roughly equivalent in their gross image-forming abilities, there are small but important differences. If we substitute a sphere for a paraboloid in a Newtonian telescope, we obtain images that look soft and mushy. The outer zones of a spherical mirror are higher than those of a paraboloid, and the angle they make with the incoming light is ever-so-slightly steeper. The outer rays, therefore, travel a shorter distance and arrive at focus too early to combine in phase properly; i.e., they come to focus closer to the objective than the central rays do. The image is larger and less concentrated than it should be, and the objective's ability to resolve stars is lessened. The image "error," or aberration, that results is called spherical aberration.

If the phase error is only a small fraction of a wavelength of light, spherical aberration will be unnoticeably small. But how little is "unnoticeably" small? A spherical mirror with focal ratio, N, and diameter, D, in millimeters, will have a ¼-wave error, a noticeably damaging amount of spherical aberration:

$$N = 1.52 \, D^{\frac{1}{3}} \quad \text{(D in millimeters)}$$
$$N = 4.47 \, D^{\frac{1}{3}} \quad \text{(D in inches)}$$

For 6″ f/8 mirrors, this lower limit is just over f/8.1. If you grind your own mirror, be sure to test it on star images before you parabolize it. The star images will look a bit unsharp. After parabolizing, they will look much crisper. This should convince you 1) that "¼-wave" does not mean "optically perfect" and 2) that the time and effort of parabolizing was well spent.

If we ask that there be no error larger than ⅛-wave, we have the following criteria:

$$N = 1.92 \, D^{\frac{1}{3}} \quad \text{(D in millimeters)}$$
$$N = 5.64 \, D^{\frac{1}{3}} \quad \text{(D in inches)}$$

The stringent requirement of 1/16-wave error, probably better than most beginners can hope to figure a paraboloidal mirror, is:

$$N = 2.52 \, D^{\frac{1}{3}} \quad \text{(D in millimeters)}$$
$$N = 7.40 \, D^{\frac{1}{3}} \quad \text{(D in inches)}$$

For 6″ mirrors, ¼-, ⅛-, and 1/16-wave accuracies require that the mirror be at least f/8.1, f/10.2, and f/13.4 respectively. Experienced telescope builders generally agree that 6″ f/12 mirrors give excellent images when left spherical, and 6″ f/10 mirrors give very good images when left spherical. This implies that 1/10-wave of spherical aberration is acceptable.

Spherical aberration results from a defect in the figure of the objective and affects the images in all parts of the field equally. Other errors of figure—uneven rings, zones, lumps, ripples, as well as smooth departures from the paraboloidal surface—obviously hurt an objective's ability to combine photons in phase. But even a perfect objective can fail if photons do not strike squarely on the optical axis.

Of the off-axis aberrations, *coma* and *astigmatism* are the two you're most likely to encounter. *Coma* is a "hairy" tail of stray light that comes about because the outer zones of a paraboloidal mirror have a slightly longer focal length than the center. Away from the optical axis, therefore, the outer zones produce an oversized image that lies outside the image formed by the center. The comatic tail always lies farther from the axis than the central part of the image.

Coma is troublesome in short focal ratio mirrors. The diameter of the region within the coma produces aberration less than a ¼-wave is:

$$CFF = 0.011 \, N^3 \quad \text{(CFF in millimeters)}$$
$$CFF = 0.00043 \, N^3 \quad \text{(CFF in inches)}$$

where N is the focal ratio and CFF the coma-free

field. For an f/8 paraboloid, the coma-free field is less than ¼″ in diameter, but for an f/5 mirror, it is only ¹⁄₁₈″. Good alignment is critical for both mirrors, but it is clearly more critical for short-focus mirrors.

Astigmatism results when rays from the same zone in a mirror do not come to focus in the same plane. As with coma, this occurs when light strikes the mirror off the optical axis. Coma is the dominant off-axis aberration in short focus mirrors. In long focus mirrors, coma and astigmatism both degrade off-axis performance.

HOW REFRACTING TELESCOPES WORK

Mirrors reflect light of all wavelengths alike. The image formed by reflectors, therefore, does not depend on the wavelength of the light. Refractors work by bending light—and, most unfortunately, light of different wavelengths is bent different as it goes through a lens. Chromatic aberration is the major aberration an optical designer must control in a refractor, thus a discussion of how refracting telescopes work is also a discussion of chromatic aberration. Typical crown BSC-2 glass, for example, has a refractive index of 1.5170 for the orange D wavelength of sodium light but indices of 1.5226 for the blue F wavelength of hydrogen light, and 1.5146 for the red C wavelength of hydrogen light. Note that the refractive index is greater for blue light and less for red light. Let's see what this means. The focal length of a simple "thin" lens element is given by the lensmaker's formula:

$$1/f = (n - 1)(1/r_1 - 1/r_2)$$

where n is the refractive index of the glass, r_1 the radius of the first surface, and r_2 the radius of the second surface. Our simple lens, therefore, will focus different colors of light at different distances from the lens.

Suppose we wish to make a lens with a focal length of 90″. What curves must we grind on its surfaces? We'll choose yellow sodium light, and for simplicity, make r_2 flat so that $1/r_2$ equals zero. The formula thus gives $r_1 = 46.53″$. We specify the lens thus:

Simple Thin Lens

$$r_1 = +46.53″$$

BSC-2

$$r_2 = plano$$

Effective focal length = 90″

A simple lens forms an image by bending the rays of light to a focus, but different colors of light have different focal lengths.

blue

red

FOCAL LENGTH vs WAVELENGTH FOR A 6″ f/15 SIMPLE LENS

Color	Wavelength	Index (n)	Focus	Error	ΔF/F
deep red	A 7665	1.5118	90.91″	+0.91″	+0.0102
red	C 6563	1.5146	90.42″	+0.42″	+0.0047
yellow	D 5893	1.5170	90.00″	0.00″	0.0000
blue	F 4861	1.5226	89.04″	−0.96″	−0.0107
violet	h 4047	1.5304	87.73″	−2.27″	−0.0253

But don't forget chromatic aberration. Above are the focal lengths of the lens at five wavelengths:

The image formed in yellow-orange light has a blue halo of light that has already passed through focus and red light that has not yet reached focus, resulting in a purple haze. If chromatic aberration were a small effect, it would not matter, but for ordinary glass, the smallest blur of light that the lens forms is around $\frac{1}{50}$ of its diameter. A star looks like a fuzzy multicolored blob seen through a telescope with a simple objective lens.

Yet we can make such a flawed lens work acceptably by increasing its focal length greatly. The color blur stays the same size, but the size of the Airy diffraction disk grows with increasing focal length and eventually exceeds the size of the chromatic blur. For a 1″ aperture, the lens must have a focal length of at least 40″; for a 1.5″ aperture, a 90″ focal length; for a 2″ aperture, 160″; for 4″ aperture, a focal length of 640″; and for a 6″ aperture, a focal length of 1440″, or 120 feet! Although much early astronomy was accomplished with lenses of such large focal length, their potential was limited. The Danzig astronomer Johannes Hevelius

built a telescope 150 feet long, but it was very difficult to use.

Achromatic lenses might have been discovered much sooner had the great English physicist Isaac Newton not made an error. Newton was one of the first to investigate refraction, and he stated, on the basis of too few measurements, that variation in the refractive index with the wavelength is proportional to the refractive index of the glass. Newton was slightly wrong; variation in the refractive index is *not quite* proportional to the refracting power. Newton went on to develop the reflecting telescope, and his pronouncement that a color-free lens was impossible to produce deterred others from trying to make one for about a century.

At the heart of the problem is glass. Below are the refractive indices for two common types of glass, crown and flint, over the range of wavelengths visible to the human eye:

Note that flint has a higher refractive index than crown glass at every wavelength. This means that if we made indentical lenses from these two glasses, the flint lens would have a shorter focal length. Note also that the

REFRACTIVE INDICES OF TYPICAL OPTICAL GLASS

Color	violet	blue	yellow	red	deep red	V
Designation	h	F	D	C	A	-
Wavelength(Å)	4047	4861	5893	6563	7665	-
BSC-2 crown	1.5304	1.5226	1.5170	1.5146	1.5118	64.5
DF-2 flint	1.6474	1.6290	1.6170	1.6122	1.6068	36.6

Modern version of a seventeenth-century long-focus refractor in use. The aperture of the lens is 2.2″; its focal length is 168″.

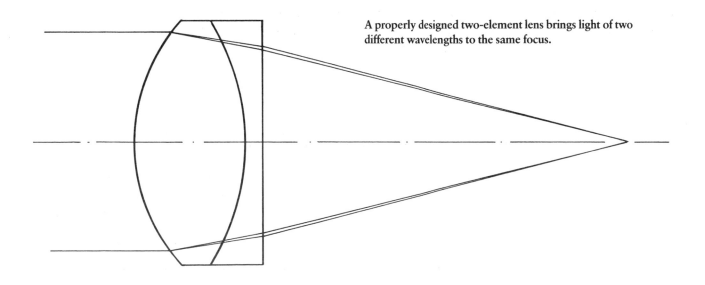

A properly designed two-element lens brings light of two different wavelengths to the same focus.

The lens elements of an astronomical achromatic refractor lens.

rate at which the refractive index changes with wavelength is greater at short wavelengths for both types of glass. This means that the chromatic aberration is more difficult to correct for violet and blue than for orange and red.

Over the range of the greatest sensitivity of the human eye, from F to C, the difference in the refractive index is 0.0080 for the crown and 0.0168 for the flint. The spread of focal lengths for a lens made of flint would be greater than for a lens made of crown, i.e., the dispersion of the light would be greater. Dividing the refractive index of D light by the difference between C and F refractive indices gives the "reciprocal relative dispersion," or the "Abbe V-number." The larger the V-number, the more light-bending power for a given amount of color spread. BSC-2 crown glass has a V-number of 64.5; DF-2 flint has a V-number of 36.6.

What does V-number mean for lensmakers? The variation in focal length between the C focus and the F focus (i.e., the longitudinal chromatic aberration) is equal to the focal length of the lens divided by the V-number. On this basis, the simple lens we computed earlier should have a focal-length difference of $90''/64.5$, or $1.4''$, between the C and F foci—as we found above.

Now, suppose you want a color-free lens. Begin by setting the condition of achromatism: The sum of the longitudinal chromatic aberration between C and F must equal zero. This is:

$$f_1/V_1 + f_2/V_2 = 0$$

Substitute the actual Abbe V-numbers:

$$f_1 = -(36.6/64.5)f_2$$
$$f_1 = -.5674\,f_2$$

This says that one of the elements must have a negative focal length and the other must have a positive focal length. The formula for combining two "thin" lenses of focal length f_1 and f_2 to make a combination lens of focal length F is:

$$1/F = 1/f_1 + 1/f_2$$

Substituting the focal length of the combination and the ratio of their focal lengths:

$$1/90'' = 1/(-.5674)f_2 + 1/f_2$$

Solving first for f_2 and then for f_1:

$$f_2 = -68.61''; f_1 = 38.93''$$

Any pair of BSC-2 crown- and DF-2 flint-glass lenses made to these focal lengths results in an achromatic lens with a focal length of 90″.

Next, calculate a set of curves for these lenses, calling them, from front to back, r_1, r_2, r_3, and r_4. We'll put the crown glass in front because it is more resistant and harder than flint. It has curves r_1 and r_2, and the flint, curves r_3 and r_4. Let's make the job easier by making the inside curves of the crown and flint element the same ($r_2 = r_3$) and giving the flint a flat back side ($1/r_4 = 0$). We now solve for curves of the lens elements by applying the lensmaker's formula:

$$1/f = (n-1)(1/r_1 - 1/r_2)$$

Solving for r_3:

$$1/(-68.61″) = (1.617 - 1)(1/r_3 - 0)$$

$$r_3 = r_2 = -42.33″$$

The negative sign means the surface is concave toward incident light; positive means it's convex toward incident light. Next, we find r_1:

$$1/38.93 = (1.5170 - 1)(1/r_1 - (1/(-42.33″)))$$

$$r_1 = 38.38″$$

The achromatic lens has the following specifications:

Flat-Back, Thin-Lens Achromat

$$r_1 = 38.38″$$

BSC-2 glass

$$r_2 = -42.33″$$

$$r_3 = -42.33″$$

DF-2 glass

$$r_4 = \text{plano}$$

Effective Focal Length = 90″

A lens designer interested in producing a top-notch lens would eliminate spherical aberration by "bending" the lens, i.e., distributing the refractive powers differently; he would also reduce or eliminate coma and astigmatism. He might use other glass types, BK-7 and LF-4 being the most common. Since there are many types of achromatic lenses, each with its particular advantages and disadvantages, the lens's use must be considered. The most suitable for visual astronomical work is the Fraunhofer objective, a coma-corrected design.

FRAUNHOFER ACHROMATIC DOUBLET

	$r_1 = 54.75″$	
BSC-2 glass		$t_1 = 0.634″$
	$r_2 = -31.90″$	
Airspace		$t_2 = 0.001″$
	$r_3 = -32.27″$	
DF-2 glass		$t_3 = 0.453″$
	$r_4 = -134.4″$	

Effective Focal Length = 90″

How achromatic *is* an achromat? We know that the C and F wavelengths focus at the same distance from the objective, but where do other wavelengths focus? A 6″ f/15 achromat typically has the following longitudinal chromatic aberration:

CHROMATIC ABERRATION OF AN ACHROMATIC DOUBLET

	Wavelength	Focus	Error	ΔF/F
A	7665	90.12″	+0.12″	+0.00137
C	6563	90.00″	0.00″	0.00000
D	5893	89.96″	−0.04″	−0.00040
F	4861	90.00″	0.00″	0.00000
h	4047	90.19″	+0.19″	+0.00210

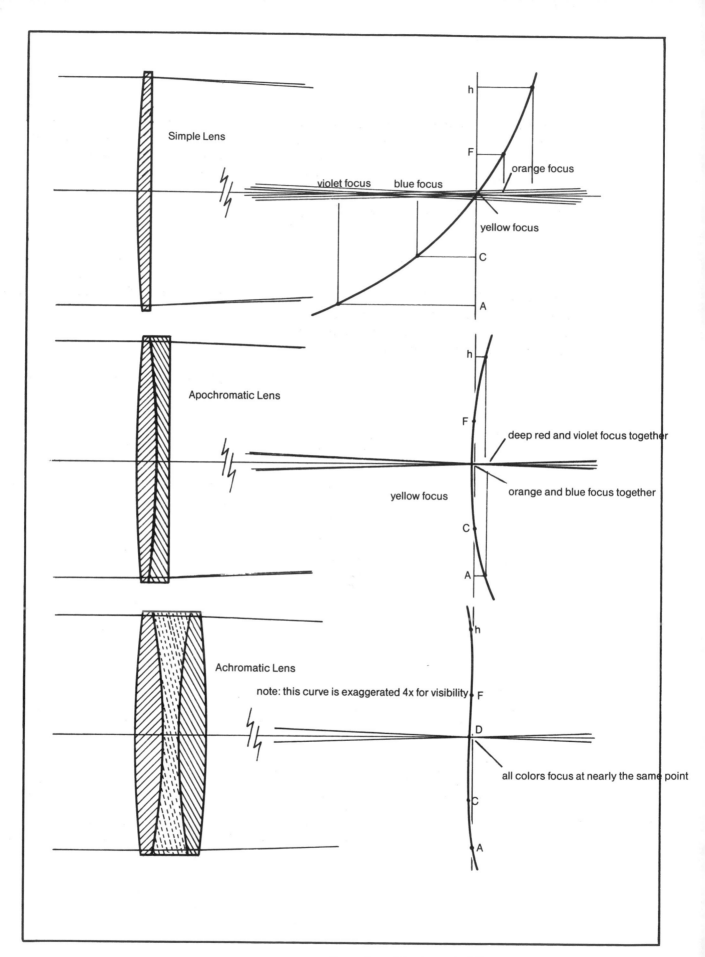

Comparison of the chromatic difference of focus for simple, achromatic, and apochromatic lenses of the same aperture and focal length.

"Secondary spectrum" is about 5 percent as great as the "primary spectrum" of a simple lens. The focal error grows considerably more rapidly in the blue than in the red, but both blue and red are evident as an out-of-focus purplish halo around the image of a bright star or planet. The generally accepted formula for the minimum focal length necessary for acceptable chromatic aberration is:

$$F_{min} = 5A^2$$

where A is the aperture in inches. Note that a 6″ f/15 achromatic objective fails this criterion by a factor of two. If this makes achromatic lenses seem hopelessly bad, remember that the sensitivity of the eye falls very rapidly from its maximum at 5500 angstroms and that the out-of-focus light is spread over a large area and for most purposes is quite invisible. In spite of secondary spectrum, an achromatic refractor with these characteristics often outresolves a reflector of the same or a somewhat greater aperture.

Apochromatic lenses reduce secondary spectrum beyond what is possible for a doublet. For an apochromatic, we must add another lens element, but the third glass must differ markedly from the first two in order to achieve the joint focus of three or four colors. The first two glasses are usually an ordinary crown and flint chosen for their stability and cost. The third glass must show an abnormal relationship between refractive index and dispersion. One such material, a special flint glass called KzFS-1, has a flintlike index of 1.613 but a crownlike V-number of 44.3. This glass is difficult to obtain, expensive, and prone to surface attack by water, but other suitable glasses have even worse properties.

The design procedure is similar to that for a doublet but far more complex since three glasses and six surfaces must be simultaneously corrected for color and other aberrations. One such lens has these specifications:

SAMPLE APOCHROMATIC LENS*

BK-7 glass	$r_1 = 92.22''$	$t_1 =$	$0.99''$
Oil	$r_2 = -23.22''$	$t_2 =$	0
KzFS-1 glass	$r_3 = -23.22''$	$t_3 =$	$0.55''$
Oil	$r_4 = +23.22''$	$t_4 =$	0
BaF-10 glass	$r_5 = +23.22''$	$t_5 =$	$0.99''$
	$r_6 = -92.22''$		

Effective Focal Length = 90″

* For demonstration only. Do not attempt to construct a lens from these numbers since the formula depends strongly on the individual batch of glass.

Such a lens has negligible color aberration. The total color error, $\Delta F/F$, over the entire visible spectrum is less than 0.0001. The overall color correction exceeds that of a typical doublet by five to ten times.

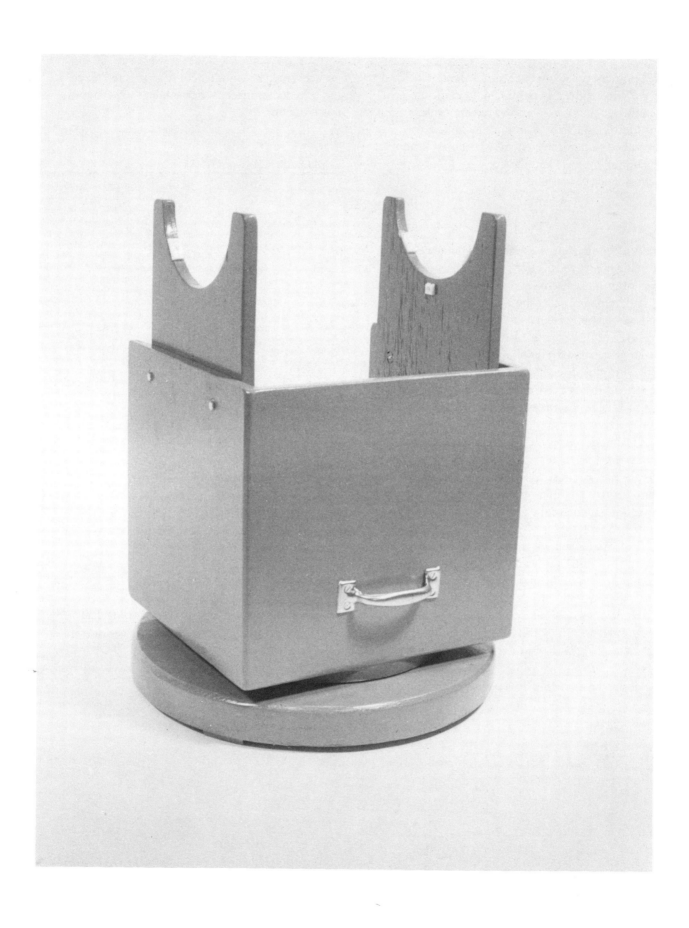

The mounting of your telescope is as important as its optical system. You should be able to point your telescope toward whatever celestial object you want to examine and enjoy a steady, unshaking "picture" of planets, stars, and galaxies even at high magnification.

Supporting, pointing, tracking, and indexing the telescopes are the functions of the mounting, but all you absolutely require for your first telescope are good, steady support and easy pointing. Later, if your interest grows into astrophotography, you'll need tracking, and for making brightness estimates of variable stars, you'll want a mount capable of indexing. Remember that you can build your telescope to go onto several different mounts for an unusually versatile instrument.

Support is the mount's main job. Without good support, none of the other functions matter. But support means a good deal more than merely holding the telescope. A telescope is a powerful optical lever magnifying the slightest changes in the angular position of the telescope. To be effective, the support structure must be rigid and free of vibrations and oscillations.

Pointing is the second important function of the mounting. A good mounting is a practical compromise between ease in pointing the telescope and stability. In the course of history, hundreds of telescope-pointing schemes have been tried. In the 1600s, astronomers hung their long refractors with ropes from a tall pole; in the 1700s, they tried great frameworks with slings and pulleys. In the 1800s, Lord Rosse's great telescope hung from chains between heavy masonry walls. Emerging from these early efforts, and fully exploited in this century, are many different kinds of mountings, each with its own advantages and disadvantages, to choose from. Underlying the diversity, however, there are only two pointing schemes: alt-azimuth and equatorial which we'll discuss in detail later.

Tracking may be necessity or simply convenience, depending entirely on the use of the telescope. Since Earth rotates, the stars appear to move across the sky. By turning the telescope, you cancel the rotation of Earth, and a celestial object, therefore, stays centered in the field. However, for many types of observation, there is no need to hold an object automatically. You can simply nudge the telescope along or turn a "slow-motion" knob to follow the sky's motion.

Indexing means "setting" the telescope on an object by its celestial coordinates by means of protractor-like

Telescope Mountings
3

circles called setting circles. To find celestial objects when you know only their coordinates, or when the observational program calls for finding many celestial objects in a short time, it's a handy feature.

SUPPORT: ELIMINATING VIBRATION

When you use a telescope, you touch it, turn the focus knob, push it, let it go. You bump the eyepiece, recenter the object you're watching, change to another eyepiece—and each contact exerts a force on the telescope. Puffs of wind may also push on the telescope, and your footsteps on the gound may shake the mount. These forces bend the structures that hold the telescope like a spring, storing energy. When let go, the mount rebounds and oscillates, in regular, harmonic motion.

The oscillation may be quite small, only a few hundredths or thousandths of a degree, but the telescope magnifies the motion. At several hundred times magnification, celestial objects in the field of view may appear to bounce around wildly. If the mounting is poorly damped, the oscillations continue for half a minute or more, and chances are that after a few

frustrating nights with a shaky telescope, you'll wish you'd never heard of astronomy!

How do we deal with the vibration? The usual approach is twofold—build a massive mount, and build it from strong materials, such as steel, brass, or aluminum. Yet telescopes built in this way are often nearly as bouncy as telescopes constructed from far lighter materials, and they have the added drawbacks of greater weight and cost. Mass and metal alone are not enough. Metals are flexible and store vibrational energy efficiently. And sheer mass, unless carefully distributed, can be a detriment.

Vibration is not a materials problem, but a design problem. Whatever the material you choose, you must design a structure that is difficult to deform, and design it to absorb energy. If it is difficult to deform, the mount will absorb energy poorly so that any oscillations induced in it start small. By absorbing energy, the mount stops vibrating quickly. The most direct route to building a damped mounting is to design in friction, friction between parts and friction in materials.

In preventing vibration from starting and in damping it out if it does, you'll bring the vibration problem under control. You can build effective mountings from a wide variety of materials that are easy to work with—

A well-damped telescope mounting vibrates only a short time when bumped.

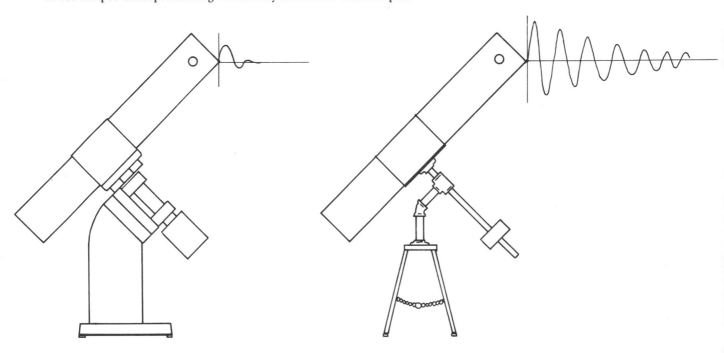

plywood, plastics, cardboard—to provide stability comparable to much heavier all-metal mountings.

The principles are simple. First, use triangles. Triangles are rigid. Squares warp into parallelograms and circles can get squashed. If you "triangulate" a structure, it can bend only if the sides of the triangles either bend or stretch. Every time you see a right-angle joint, form a triangle with a brace. Every time you support something from one side or hang it off to one side, you've built a "weight-on-a-stick" oscillator. Support telescopes from both sides whenever possible.

Second, design your mounts to absorb energy. Without internal friction, there is no way for the energy of oscillation to travel from the shaking telescope to the stationary mounting. Choose bearings that have large areas of contact. And don't worry about the friction in the bearings. Telescopes turn so slowly that the power needed to overcome friction is negligible. With precision, low-friction bearings, you'd probably need to *add* friction with a clutch plate in the drive to keep the telescope from turning too freely.

Divide vibration between different modes of oscillation. If the telescope can oscillate in a multitude of ways, none will be large. Avoid simple structures (piers, rods, pillars), and brace beams internally so that each section vibrates in its own way.

Finally, don't be afraid of "soft" materials in a telescope. With metals, the mounting has no energy sinks, and oscillations can continue for a long time. Use materials with internal friction; plywood, plastic, and paper tube soak up oscillation.

POINTING: BEARINGS FOR TELESCOPES

To locate any place in the sky, the observer rotates the telescope about two mutually perpendicular axes. Every telescope mounting, whatever its type, must have two mutually perpendicular bearings. Pointing is then a matter of rotating the telescope about these axes until it is aimed in the desired direction. In virtually all telescopes, one bearing carries the telescope tube directly and another holds both the first bearing and the telescope. When the main bearing turns, both the telescope and the other bearing rotate with it.

The two major classes of telescope mountings

correspond to the two coordinate systems in common use in astronomy: the altitude-azimuth system, oriented relative to Earth's surface; and the equatorial system, oriented relative to Earth's spin axis.

Regardless of the class or type of mounting, however, all bearings perform the same function: They restrict motion. Any body, whether it's a brick or a telescope, can undergo only six types of motion: three independent translations (up/down, back/forth, and left/right) and three independent rotations (roll, pitch, and yaw). All other motions can be broken down into combinations of these six basic "degrees of freedom."

In a telescope mount, each bearing consists of an axle held by two restraints. Each restraint is what you probably think of as a "bearing"—a ball race, a bushing, a vee block, tight-fitting rings, sliding surfaces, rollers, or some other device to "bear" a load. Ideally, each bearing would allow only rotation, but practically, all of them allow some unwanted motion. By supporting each axle with two restraints, we can all but eliminate this. You may then point a telescope toward any place on the celestial sphere by mounting it on two mutually perpendicular bearings.

The five basic bearing geometries—shaft, conic, equilateral, pin and disk, and plate—are functionally the same. They may be combined in virtually any way imaginable, producing scores of different kinds of mountings: You can enlarge your mounting to any size, cut parts of it away, or reduce it to a skeleton framework—all without disturbing the basic character of the bearing.

Furthermore, the telescope can ride anywhere; if it lies between the restraints, it is "inboard;" if cantilevered outside them, it is "outboard." Mountings that carry the load inboard are almost invariably more rigid and less prone to vibration, but outboard mounts offer greater convenience and easier access to the telescope.

You are probably the most familiar with shaft bearings: an axle supported by two identical restraints, usually ball races. Shafts have a few problems. Attaching a telescope to the end of a metal bar is not easy, and the bar itself may be flexible. One solution is to enlarge the whole shaft to make it more rigid and provide a larger attachment surface; better is enlarging only the upper end so the shaft becomes a cone. This allows it to bear the weight of the telescope with less flexure and still provide a wider base for attaching the telescope.

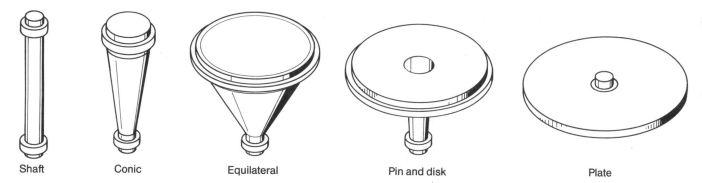

| Shaft | Conic | Equilateral | Pin and disk | Plate |

The five basic bearing geometries.

The bearing at the large end of the conic shaft can carry much of the weight, while the small end carries a lesser share and prevents unwanted motion.

Carrying the evolution of the axle still further, we enlarge the big end of a conic shaft until its diameter equals the length of the shaft between the supports. The axle now consists of a fat, or equilateral, cone. The large end can rotate in a large ball race or bear on rollers; the small end prevents the cone from tipping. Such a bearing is well suited for carrying lots of weight. Imagine enlarging this bearing until it becomes larger than the telescope itself, then cut away the center and drop the telescope inside it. You'll end up with the mounting of the 200″ Hale telescope on Mount Palomar.

The pin and plate is a further development of the equilateral bearing: The large end is a disk bearing most of the load, and the shaft is a mere "pin" to stabilize the disk and prevent it from overturning. Allow the pin to shrink until it disappears into the center of the disk, transforming the axle into a flat plate. Its function is identical to other bearings in allowing just one rotational degree of freedom, but it has come to look very different. All support now comes from rollers or sliding contacts around or behind the plate. Plate bearings are exceptionally stable right-ascension axes for equatorial telescopes, or azimuth bearings for alt-azimuth telescopes.

ALTITUDE-AZIMUTH MOUNTINGS

A few decades ago, all large telescopes were mounted on equatorial mountings. But as professional astronomers built telescopes with greater light-gathering power, these giants became progressively more unwieldy and expensive to mount. They finally gave way to the steadier and less expensive alt-azimuth type of mounting in the early 1970s, when two of the world's three largest telescopes, the Soviet 6-meter reflector and the 4.2-meter Multiple Mirror Telescope, were designed with alt-azimuth mounts. Because of the success of these pioneering instruments, it is unlikely that astronomers will again choose an equatorial mount for a world-class professional telescope of over 4 meters' aperture. The alt-azimuth mounting offers stability at much lower cost and can carry the load of a big telescope with less flexure. Computer-controlled, variable-speed-drive motors on both axes are less expensive than mechanical structure of a suitable equatorial mounting.

Amateur astronomers rediscovered alt-azimuth mountings in the late 1970s for essentially the same reasons: Equatorially mounted telescopes are expensive and prone to vibration. Modern alt-azimuth mounts (the Dobsonian mount in particular) combine stability, low cost, and ease of construction. The slick fluorocarbon plastic used in their bearings

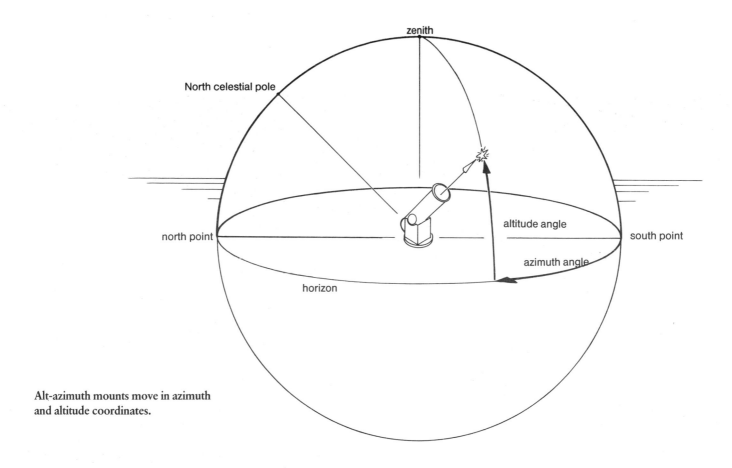

zenith

North celestial pole

altitude angle

north point

south point

azimuth angle

horizon

Alt-azimuth mounts move in azimuth
and altitude coordinates.

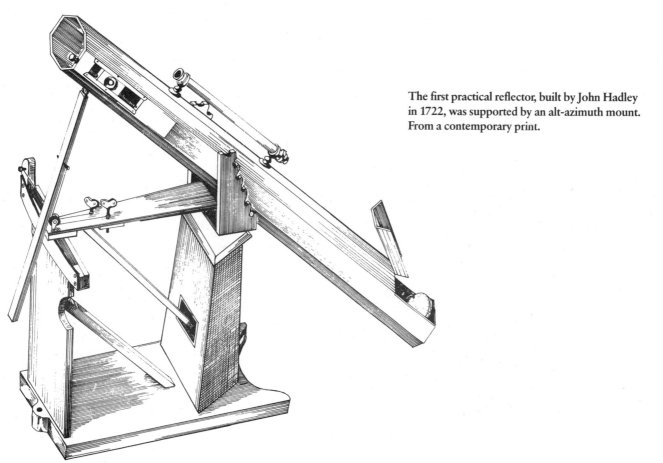

The first practical reflector, built by John Hadley
in 1722, was supported by an alt-azimuth mount.
From a contemporary print.

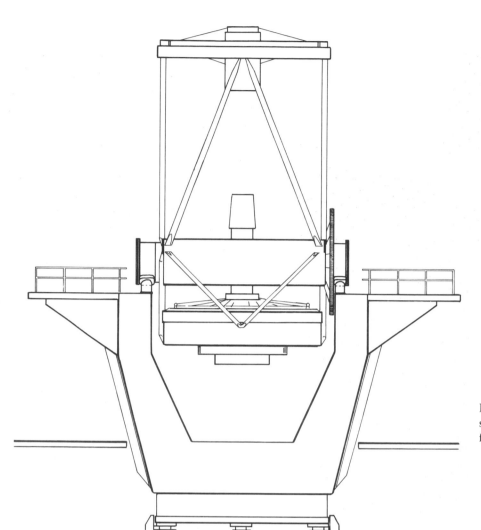

Britain's 4.2-meter William Herschel telescope rides an alt-azimuth
fork mounting.

makes smooth "hand-and-eye" tracking acceptable
for many kinds of observing.

Hadley's mount has been with us since the 1720s,
when John Hadley mounted the first practical Newtonian telescope, a 6″ f/10.3, between the arms of a box
that turned in azimuth. In most versions of Hadley's
mount, an outboard shaft-type azimuth axis supports
the telescope on an inboard shaft. An adjustable rod
steadies the telescope in altitude. In Hadley's original
mount, a key-and-pulley system controlled both the
azimuth and the altitude motions.

However, Hadley's mount and modern variations of it,
while not cumbersome, are not especially convenient. In
order to follow a star, the observer must continuously
adjust the pointing of the telescope in both axes. This is
either a two-handed job or requires alternating one hand

between the two controls. If you're serious about using
the telescope, you'll soon master the observing technique,
but it doesn't come easily or naturally.

The alt-azimuth fork is most often supplied with small,
imported refractors. In its least-workable versions, this
mounting has given alt-azimuth mountings a bad name.
The azimuth bearing is usually a very short conic shaft
bearing, and the altitude is an inboard shaft bearing, all
placed atop a wobbly tripod. Since there is often too
little friction to hold the telescope steady in either axis,
the axes are often supplied with a "steadying rod,"
which causes the telescope to shift when it is clamped.
As with Hadley's mount, the pointing in the two axes
must be corrected independently.

Alt-azimuth forks are not inherently bad, however.
Late nineteenth-century refractors—a pleasure to use—

were mounted on them. They usually had large conic-shaft azimuth bearings, adjustable-tension altitude bearings, and a heavy brass fork atop a sturdily braced wooden tripod. Many of these instruments are still around. Their bearings have enough friction that the telescope remains where you leave it but little enough friction that an intentional nudge on the tube will move the telescope to follow a celestial object.

Professional-sized telescopes such as the Soviet 6-meter, the British William Herschel Telescope, and the Multiple Mirror Telescope are on alt-az forks. The azimuth bearings of both are large-diameter equilateral bearings, and the altitude bearings are massive shaft-type bearings that carry the telescope between the arms of the fork. Both axes are driven by variable-speed motors controlled by a digital computer.

Dobsonian mountings are by far the best of the alt-azimuth mounts for home-built telescopes. The Dobsonian is a variation of the alt-az fork mount adapted for use with Newtonian reflectors and designed for construction with low-cost materials in mind. The azimuth axis is a large disk bearing made of hard plastic sliding smoothly on slick fluorocarbons pads. This bearing carries a box that supports the large-diameter inboard bearing carrying the telescope.

The large cross sections and contact surfaces mean that the Dobsonian mounting is exceptionally stable. Vibrations induced in either tube or mount rapidly damp out. Since both bearings have a significant amount of internal friction, you must apply small, deliberate pressure to move the telescope. The telescope remains pointed at any part of the sky without clamps or stabilizers but responds to a light touch.

The feature that truly distinguishes a Dobsonian mounting, however, is that *the telescope moves simultaneously in both axes* when pushed. There are no clamps or knobs; when you nudge the tube, it moves in the direction in which you nudge it. The instant you

The Dobsonian mount is stable, practical, and compact. Telescope by Earl Watts.

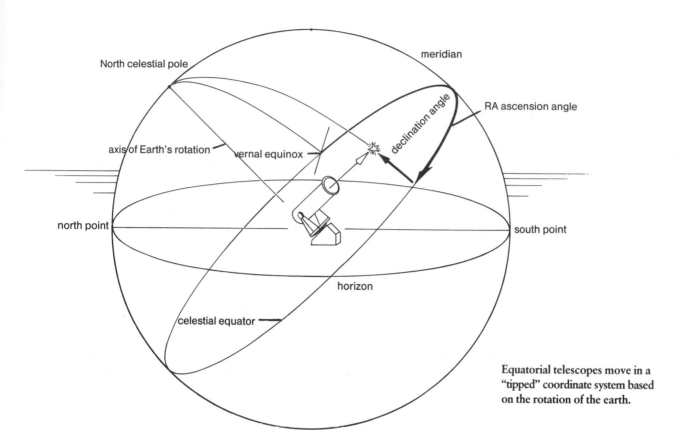

North celestial pole

meridian

RA ascension angle

declination angle

axis of Earth's rotation

vernal equinox

north point

south point

horizon

celestial equator

Equatorial telescopes move in a "tipped" coordinate system based on the rotation of the earth.

stop nudging, the telescope stops. For *solid* support and *easy* pointing, the Dobsonian mount is nearly ideal.

EQUATORIAL MOUNTINGS

Since their bearings rotate parallel to Earth's spin, an equatorial mount makes tracking and indexing mechanically simple. An equatorial mount with a drive "freezes" the motion of the sky; stars remain fixed in the field of the eyepiece or on a photographic emulsion. Indexed setting circles allow you to dial in astronomical objects, which is valuable for serious observing programs. But much observing, especially visual observing, does not require an equatorial mounting.

When the question of mounts comes up, consider your observing program. For visual observing, a shaky equatorial is considerably less useful than a solid alt-

azimuth, and it's doubtful that you'll be happy with a shaky equatorial for any kind of observing. Therefore, if your observing program requires an equatorial, be prepared to build a first-class one and nothing less.

There are a variety of equatorial mountings to choose from. All carry out tracking and indexing nearly identically since these two functions depend primarily on the diameter and quality of the drive mechanism and circles. When you come to make a choice between equatorials, consider your needs for portability and weight, the stability of each, and the ease with which you can reach the eyepiece of the telescope.

The German mount is the most common type of equatorial for home-built reflecting telescopes and just about the only type used for equatorial refractors. Unfortunately, both axes carry the load of the telescope cantilevered beyond the support points, creating a vibration problem that is hard to cure. The greatest drawback is that the telescope cannot cross the meridian without hitting the pier. When the telescope reaches the meridian, it must be "flipped" over the pier to the other side and spun 180° in declination. This is confusing for beginners and a nuisance in photography.

However, German mounts are not unsuitable for home-built telescopes. You need not use a shaft polar bearing, for example, but may use the plate or pin-and-disk style. The disk or plate effectively transmits telescope vibrations down the pier to the ground, and the declination axis supports are wide and thus relatively stable. They are portable, but clumsy to pick up and carry. While access to the eyepiece is good, a rotatable tube is a necessity because of the pier problem.

The yoke, or English mount, is the antithesis of the German mount: Whereas the German is doubly outboard, the yoke is doubly inboard. Because of its design, the yoke mount is inherently stable. The telescope rides be-

tween the supports of the declination axis, and the yoke assembly, an enlarged and split shaft, rides between the polar axis supports. Because yoke mountings are probably the most easily built of any of the extremely stable doubly inboard equatorials, astronomers chose a yoke for the 100″ Mount Wilson reflector, the first of the twentieth century's giant telescopes.

Yokes permit tracking across the meridian, require no counterweights, and carry heavy telescopes without trouble. They are thus suitable for deep-sky astro-photography. However, the eyepiece is hard to reach if the tube does not rotate; the sky near the pole is blocked by the north polar bearing; and yokes are not portable mounts. If you want a solid, nonportable mount that is easy to build, and if you plan observing that doesn't demand access to the sky near the pole, a yoke is an excellent choice.

German equatorials have long been popular among amateur telescope-makers. Telescope by Bob Webb.

The **cross-axis** is a close relative of the yoke mounting and is sometimes also called an English mounting. The polar axis is an enlarged inboard shaft bearing, and the declination axis an outboard bearing, usually a shaft or tapered shaft. Many of the 36″- to 84″-aperture professional telescopes built in the 1920s and 1930s ride on cross-axis mounts.

Like yoke mounts, cross-axis mounts are extremely stable. They also track across the meridian without interruption and can carry heavy telescopes. The eyepiece is relatively accessible, and the mount permits pointing the telescope at any part of the sky. On the negative side, the cross-axis mount requires a counterweight, and it is seldom portable because of the massive polar axis and its supporting piers. For a permanent installation, the cross-axis is exceptionally well-matched to the requirements of deep-sky astrophotography or photoelectric photometry.

Horseshoe mounts are relatives of the yoke. All share one fundamental of the doubly inboard family: inherent stability. Horseshoe mounts are derived from the yoke, changing the relative sizes of the parts. If you transform the polar axis into a pin-and-disk bearing with a massive split pin, then cut away the parts of the disk that block the telescope's view of the pole, you'll have a horseshoe mount. The 200″ (5-meter) Hale telescope on Palomar Mountain, the twin Kitt Peak and Cerro Tololo 4-meter telescopes, and the 3.6-meter telescope at the European Southern Observatory have horseshoe mounts.

Like the rest of the yoke family, horseshoe mounts track across the meridian, require no counterweighting,

Yoke mountings are large but very solid because the telescope is fully "inboard."

Cross-axis mounts are excellent for astrophotography from a permanent observing site.

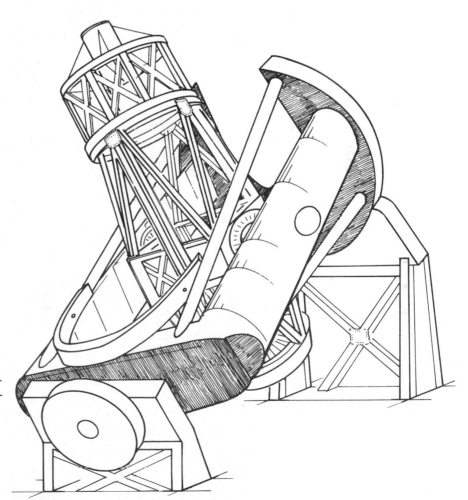

The 200″ Hale telescope on Palomar Mountain is the largest equatorially mounted telescope in the world; its mount belongs to the "horseshoe" class.

Fork mounts are suitable for relatively light home-built telescopes. Telescope by Tom Greska.

Inverted forks are excellent mounts for a first telescope. Telescope by Alan Beaman, age 8.

and can carry heavy telescopes. They are more compact than an equivalent yoke, an important factor in the design of extremely large professional telescopes where the dome may represent a significant fraction of the cost of the installation. For amateur telescopes, the mechanically simpler yoke and cross-axis mounts are superior.

Fork mountings are among the most common mounts for medium-sized professional telescopes and are popular in commercially-made amateur telescopes. Forks are outboard in the polar axis and inboard in the declination. You'll see a variety of bearing types used for the polar axis, but declination is always a shaft bearing. The "fork," after which the class is named, is the U-shaped structure supporting the declination bearing.

Although fork mounts are quite successful in supporting large professional telescopes, the fork itself is a tricky part. While it is simple to build a fork strong enough to support the weight of a telescope, eliminating its vibrations is usually difficult.

Forks provide tracking through the meridian, can point to the celestial pole, provide good observer access to the telescope, and are relatively compact. They are poorly suited to carrying heavy telescopes, and access to an eyepiece or instruments mounted on the back end of the telescope may be poor when the telescope is pointed near the pole.

The inverted-fork mount is a derivative of the German mount. As in the German, the polar axis is outboard, but, unlike the German, the declination axis is neither inboard nor outboard since the telescope rides to one side of the declination axis, balanced by counterweights. Because the width of the fork limits the diameter of the polar bearing, inverted-fork mounts usually have a shaft or tapered-shaft polar axis.

Also unlike the German mount, inverted forks track across the meridian without hitting the pier. The greatest advantage of the inverted fork is that the observer has good access to the eyepiece even if the tube of the telescope can't be rotated. An inverted-fork mount is easy to build and provides good stability for small telescopes.

"Asymmetric equatorial" describes a large number of relatively sophisticated mounts. The class includes one-armed forks, bent yokes, and the torque-tube equatorial, to name just a few. These designs try to eliminate the

One-armed fork mounts of the asymmetric equatorial class require machining skills to build. Telescope by Orien Ernest.

counterweight, or to at least tuck it out of the way. Often the declination-axis support structure serves to balance the polar axis, although in the torque-tube equatorial, the counterweight has simply been shifted down the polar axis, where it does not get in the way of the observer. Asymmetric mounts are usually designed with a specific telescope in mind and optimized for it.

TRACKING: TELESCOPE DRIVES

Equatorial mountings can be "clock-driven" to follow, or "track," a star by turning the polar axis once per sidereal day. In the nineteenth century, drives were clockwork mechanisms with mechanical governors. They worked but tended to be underpowered and finicky. Since the beginning of the twentieth century, electric motors have been used. The motor turns a train of gears that drive a large precision gear on the polar shaft.

How good should the drive be? It depends almost entirely on the reason for driving the telescope. For casual visual observing, the drive rate can vary by 1 percent, and indeed, this much error is common in commercial telescope drives, most of which are built to turn the telescope once in 24 hours—one solar day—rather than the 23.9345-hour sidereal day, a built-in error of 0.27 percent. More annoying, however, are irregularities in the drive system that cause large, short-term rate variations. While the overall speed of the motor may be corrected with a variable-frequency power supply, rapid periodic variations must be monitored and "guided" out.

To gain a feel for what a drive must do, consider observing with no drive at all. At low magnification, an object remains in the field of view long enough; at 40x magnification with a 10″ reflector, an object near the celestial equator takes four minutes to cross the field. Even with more magnification than most observers use, say 400x, an object takes about 40 seconds to cross the field. An experienced observer simply nudges the telescope every ten or fifteen seconds to recenter the object.

Now consider your requirements for serious planetary observing. Suppose you insist that an acceptable drive must maintain the object under observation within the central 10 percent of the field of view at the highest magnification field ever used, 500x, at all times. This criterion allows an angular error of about $\frac{1}{100}$°, or 36 seconds of arc, and represents an 0.1 percent cumulative error over an observing period of 40 minutes. Even though a solar-rate drive appears

Telescope gear-drive systems must have large numbers of accurately cut teeth—in this case 221—with precision-matched worms.

not to meet this criterion, you could slow the motor to sidereal rate with an electronic drive corrector. Somewhat more annoying, but still acceptable, would be a regular variation in the rate—alternately 1 percent fast and 1 percent slow—having a period of four minutes. Such errors are typical of cheap worm-gear drives and would alternately place the object 18 seconds of arc ahead and then 18 seconds of arc behind its mean position, for a total excursion of 36 seconds of arc.

Ordinary commercial gears intended for power transmission, while "precise" by ordinary standards, are barely good enough for casual visual use. The best commercial telescope drive-gear sets on the market are about one hundred times better and are suitable for long-exposure, deep-sky astrophotography. These gears have a large number of accurately cut teeth—a minimum of several hundred—and are available with matching worm and motor. The production of such gears requires elaborate machine-shop equipment and considerable skill.

Although we have called the driving device a gear, there are other approaches to making drives. Home-built telescope drives often rely on the technique of "borrowed precision." Mass-produced threaded rod, for example, has thread-to-thread errors well under 0.001″ and can be bent. To minimize short-term drive-rate errors, you need a large driving radius and must rely on a drive corrector to adjust the overall rate precisely. The driving radius should be as large as you can make it—as much as half the focal length of the telescope. Although a big drive "gear" won't fit on a portable mounting, a 30° sector can be tucked away on mountings such as the yoke and cross-axis.

INDEXING: SETTING CIRCLES

When the axes of an equatorial mounting are parallel to the axes of the coordinate system, the angular position of the telescope and the coordinates of the sky coincide. By installing indexed circles on the axes, you can "set" the telescope on a given celestial object providing you know its coordinates. Setting circles, as they are called, speed up observing because you need not consult star charts but can simply point the telescope at an object. For repetitive or routine observing, such

A clock drive and setting circles enhance the usefulness of an equatorial mount for prolonged observing sessions. German equatorial by Bill Shaefer.

as estimating the brightnesses of variable stars or searching galaxies for supernovae, setting circles are extremely useful. However, unless they let the observer locate the desired object near the center of the field every time, almost without fail, circles may be a frustrating liability.

For one thing, small circles are useless. The old rule of thumb that states that setting circles should be at least, but preferably twice, as large as the aperture of the telescope is fairly accurate.

To locate variable stars, a 10″ reflector with a field of 0.4° at 100x requires 0.1° setting accuracy. If you can reliably set the index scale to 0.02″ then a degree on the periphery of the circle should equal 0.2″, for a circle with a diameter of 22″! You may feel that such

Large-diameter setting circles are necessary for accurate telescope indexing.
Antique refractor mounting restored by Pat Michaud.

setting circles are excessively large and decide you'll chance reading the circles to 0.01″ or acquire objects with a wider field at lower magnification.

You can buy circles, make them on a machinist's dividing head, or carefully draw a scale on a strip of flexible material and wrap it around a large metal or wooden circle. Stamp or mark the complete 360° strip (plastic or metal), then adjust the radius of the circle until the ends of the strip just meet. Glue the strip in place, or hold it with small screws or nails. Declination circle divisions should be in degrees, with every fifth or tenth degree accentuated. If you subdivide the degrees, do so into 2, 3, or 6 parts to aid estimating to minutes of arc. Right-ascension circles should be labeled every hour, divided into 30-, 10-, and 5-minute intervals, and if further subdivided, into one-minute intervals.

It is possible to use circles on alt-azimuth mounts, although you must first perform the mathematical "transformation" from one coordinate system to the other. Converting from hour angle and declination to altitude and azimuth with a portable computer takes but a few seconds.

EQUATORIAL PLATFORMS

In the late 1970s, Frenchman Adrien Poncet realized that only part of an equatorial mounting is necessary for tracking. Poncet's original platform was a table supported by three points: a pivot point and two feet sliding on a plane inclined normal to the earth's polar axis. The feet constrained the table to rotate only about an axis parallel to the earth's axis.

Imagine a horseshoe mounting, one such as the 200″ Palomar telescope uses, cut off and covered with a large deck. Even after this major surgery, the lower part of the mount still turns about an axis parallel to the earth's axis of rotation. If you put a telescope on the upper part, it shares the motion of the platform and follows the stars.

Equatorial platforms enable telescopes on alt-azimuth mountings to track equatorially without the complexity of an equatorial mount. Although they are limited by the tilt of the platform to about an hour's tracking time, they make large-aperture Dobsonian telescopes

capable of deep-sky photography without the expense of a full equatorial mount.

SPECIAL MOUNTINGS

Springfield mountings are German equatorial mountings with disk-type bearings in both polar and declination axes. Their special feature is that the light path of the telescope passes into the axes of the mounting so the eyepiece remains fixed regardless of where the telescope is pointing. The mount gets its name from Springfield, Vermont, the home of Russell W. Porter, its inventor and the mentor of the telescope hobby. There are several drawbacks to the Springfield: It works best with long focal-ratio optical systems; optical alignment is difficult to maintain; and warm air from the observer rises into the optical path. Nonetheless, a Springfield mount is an excellent choice for a permanently mounted telescope, especially if your observing program calls for long stints at the eyepiece.

Turret mountings are another Porter invention. The most famous turret, built in 1930, is at Stellafane, the clubhouse of the Springfield amateur telescope-makers. Its polar axis is a giant disk bearing carrying the primary mirror of the telescope on a long arm perpendicular to the axis of rotation. A large, flat mirror rotatable in the declination axis directs light from a star to the primary, where it returns, through a hole in the flat mirror, into the turret, and to the observer. The turret can be heated inside and otherwise made comfortable. A turret telescope is a major undertaking and requires a large precision flat mirror.

Ball mountings differ from other mountings because they have no axes, but they can still be pointed easily to any part of the sky. The telescope rides in or on a sphere supported on three points and therefore has two degrees of freedom in its motion. A model of Isaac Newton's first reflector, built in the eighteenth century, is mounted on a wooden sphere about the size of a croquet ball, but ball mountings were little more than curiosities until Edmund Scientific introduced a molded plastic 4″ f/4.5 Newtonian reflector, the Astroscan, with an integral ball. Ball mounts are vibration-free, very portable, and easy to operate. Home-built ball mounts often employ a bowling ball.

Building a 4″ f/10 Reflector

4

Here's a telescope that's perfect for beginners. It won't eat up your budget, and you can put it together in about a week of evenings once the optical parts arrive. It's a great telescope if you'd like to just "stick a toe in the water," and it's simple enough for preteens to build. It's tough—in fact, just about indestructible—easy to use, and a good performer. It'll show you everything *any* telescope with a 4″ lens or mirror can. Years after building bigger and better telescopes, you'll still come back to this one for the sheer joy of using it.

The telescope is an equatorially mounted Newtonian reflector. The mount is an inverted fork of very simple design. The telescope tube is a wooden box. Its mechanical parts are made of pipe, plywood, and hardware that you can buy at just about any hardware store. The optical parts—the primary mirror, the secondary mirror, the eyepiece holder, eyepieces, and possibly a finder telescope—may be purchased ready-made from mail-order firms. Even the tools needed are simple: a crosscut handsaw, a compass saw, a hand drill, screwdriver, chisel, carpenter's square, pliers, sandpaper, and paintbrush.

But heed this warning: While you're working on your telescope, start learning the stars and constellations if you don't already know them. If you're helping a son or daughter build it, or if it's a scout project, take the youngsters on a trip to the planetarium. Find out where the planets are. Get out your binoculars and start to explore the sky. Remember, a telescope can show you a lot, but only if you've learned what to look at and where to find it!

ORDERING THE OPTICAL PARTS

Begin by ordering the optical parts. Deliveries sometimes take a couple of months. Consult Appendix B, science-company catalogs, and the ads in current issues of astronomical magazines. First and foremost, you'll need to order a 4¼″ f/10 aluminized and overcoated spherical primary mirror and an elliptical diagonal mirror with a minor axis of about 1″. Some suppliers sell both mirrors as a set; others sell them separately. A few companies even include lenses that you can make a simple eyepiece with. Primary mirrors cost $30 to $40, and diagonal mirrors around $10.

This snapshot of the moon, taken through the 4″ f/10 reflector shown in this chapter, barely hints at the detail visible through the eyepiece on a crisp, clear night.

The telescope's optical parts: the primary mirror, the diagonal mirror, and an eyepiece.

Next, order a 1¼″ eyepiece holder. The 1¼″ means that the holder will hold eyepieces with barrels 1¼″ in diameter. "Helical" and "push-pull" focusers work nearly as well as the more expensive rack-and-pinion varieties, so you don't have to spend a lot. Order a diagonal support stalk with the holder. This is not the best way to hold the diagonal, but it is good enough.

If you're trying to hold down costs, don't be afraid to make your own eyepiece holder from plastic pipe or a wooden block and plastic plumbing. Ideas for this kind of focuser are in Chapter 9.

Next, order an eyepiece. Start with a low-powered eyepiece, one of about 25mm focal length. It will give a magnification of around 40x. The relatively "slow" f/10 focal ratio of the mirror will not show up the faults of eyepieces, so even a simple Ramsden will perform pretty well. The eyepiece should have a 1¼″ barrel to match the focuser. If your budget permits, you might get a second eyepiece, this one with a focal length of between 8mm and 12mm. Without getting into the hassles of high-magnification observing, a medium-powered eyepiece will show you just about everything a 4″ telescope can.

Finally, think about ordering a finder telescope. A finder is a little telescope used for aiming the main telescope. If your budget is tight, defer the purchase until later. Finders tend to be expensive; a good finder will cost more than the primary mirror! Since this telescope *can* be used with a peepsight finder wait and see how much (or how little) you need. Chapter 9 shows you how you can build a peepsight or finder yourself.

With the orders in the mail, you will have spent about half of the total cost of the telescope. The rest will be spent at the hardware store.

A TRIP TO THE HARDWARE STORE

Now you can start scouting out parts for the tube and mount. First, you'll need "precision equatorial bearings," ordinary 1½″ pipe fittings, for the mounting. You'll need a tee, two close nipples, two floor flanges, and one 12″ nipple. This size of pipe is carried in well-stocked hardware stores, but if it's not in your local outlet, try a plumbing-supply store. The pipe

will cost around $15 unless you have an uncle in the plumbing business.

While you're buying the pipe, purchase four U bolts that fit snugly around the pipe and extend an inch beyond it. It's a good idea to check this while you're still in the store. As you will see later, the U bolts hold the mount together.

Below is a list of the hardware you need. You don't have to buy it all at once.

For the Tube:
2 ¼-20x½″ tee nuts
3 10-24x½″ tee nuts
1 ¼-20x¾″ tee nuts
1 ¼-20x1¼″ hex-head bolt
3 10-24x1″ round-head bolts
2 ¼-20x1½″ thumbscrews
2 ¼-20 hex nuts
2 1″ fender washers
24 #8x¾″ flat-head wood screws
1 small box ¾″ wire brads

For the Mounting:
2 1½″ floor flanges
2 1½″ short nipples
1 12″ nipple, or 12″ pipe threaded one end
1 1½″x1½″ tee union
4 U bolts, must fit 1½″ pipe
10 ¼-20x¾″ tee nuts
10 ¼-20x1″ flat-head bolts
¼ pound 6d finishing nails
24 #8x2″ flat-head wood screws

Next, visit a lumberyard or home-improvement center and buy a 2′x4′ panel of ¼″ plywood (often sold as "handi-panels"), a 2′x4′ panel of ¾″ plywood, and 16 linear feet of ¾″ quarter-round molding. Prices vary, but you'll probably spend under $20.

LAYOUT AND CUTTING

Study the plans for the 4″ f/10 reflector. Become familiar with each part that will go into the telescope. The telescope itself, or more precisely, its tube, will be cut from the 2′x4′ piece of ¼″ plywood. This will

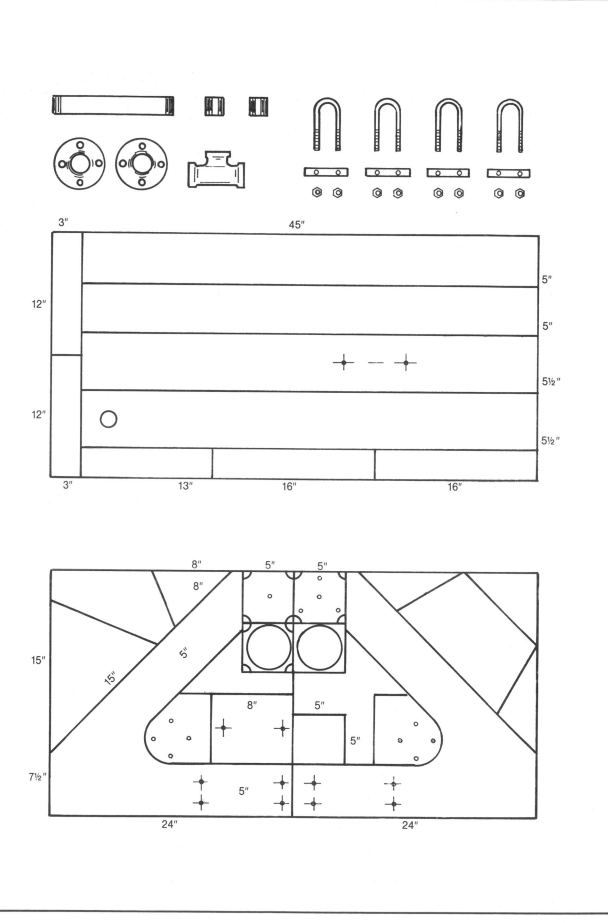

All parts for the 4″ f/10 reflector can be cut from two quarter-panels of plywood, one ¼″ thick and the other ¾″ thick.

be the sides of the tube. You will use the leftover material for the arms that hold the counterweights and to reinforce the tube. The ¾″-thick piece of plywood provides the remaining parts for the tube: the bulkhead rings at the front end and in the middle for reinforcement, the mirror cell, the big v-shaped base pieces (which we'll call the "vees"), and the fork parts.

The first step, before you cut any wood, is to transfer the entire layout to the wood. Begin with the sheet of ¼″ material. Place it on your worktable under good lights. Measure its dimensions. It probably won't be *exactly* 24″x48″, but it should be close. Carefully mark off at each end of the wood the lengths shown on the plan, using the inch marks on your carpenter's square; do the same in the middle; then draw pencil lines to connect them, using the long, straight side of the square or the side of a straight piece of lumber. Mark off the end and side pieces that will be 3″ wide, and the 16″-divisions in them. Next, double-check all the dimensions. There should be nine pieces coming from this sheet: four sides, two counterweight arms, two counterweight arm reinforcers, and the tube reinforcer.

Now do the same for the ¾″-thick sheet. Begin by drawing a line that divides it right down the middle; then measure in 4½″ from one side and draw a line parallel to the side. Measure 6″ up on each of the ends. Check to see that 18″ remain on the side; then, on the adjoining long side, measure off 18″ and connect them with a line. This is the bottom of the mount. Measure 4½″ from this line and lightly draw another line parallel to the bottom line.

Next, locate the two center points shown on the plans, and, using the square, mark them on the wood. With a large compass, draw arcs with a radius of 4½″ from the center points, intersecting the inner edges of the vees. Finally, draw two lines parallel to the line that divides the sheet and 5″ on either side of it. You now have all the lines that define the two vees. Double-check all the dimensions, and if they're right, darken the lines around each vee. Finish marking the sheet by measuring off the spreaders, the base, the fork parts, the stabilizer, and the remaining tube parts. Check again to make sure that all the dimensions are correct.

Start with the ¼″ sheet. Score the plywood along

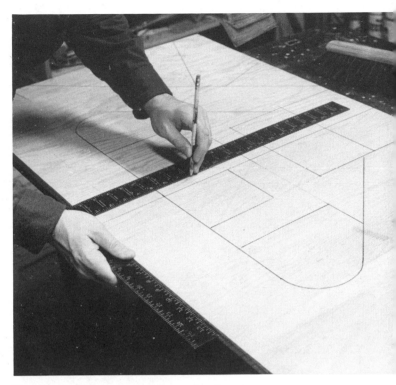

Transferring the layout to wood precedes all cutting.

Getting started at last; do the straight cuts first.

A saber saw is best for cutting the curves.

Clamp them together, then plane or surform the edges of
the vees and other paired parts until they match.

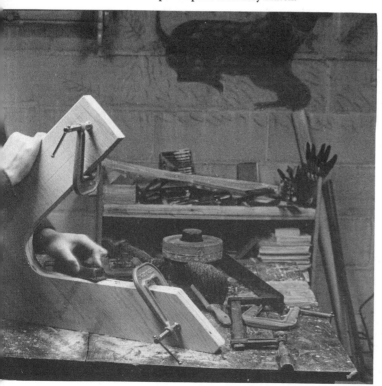

the cut with a sharp hardware knife to minimize ragged
edges. Use your hand saw for the straight cuts. First,
saw the 3″-wide end. Work steadily. Don't rush, and
you'll find it goes quite quickly. Next, saw the sheets
lengthwise for the four sides of the tube and the 3″-wide
counterweight arms.

On the ¾″ sheet, begin with the center cut; this
way, each piece will be smaller and easier to handle.
Next, cut along the vee bottom lines, removing two
large triangles, then the cell and tube parts, then the
straight segments of the inner edges of the vees. Use
a compass saw (or power saber saw) for the curved
cuts. Finish all the remaining cuts except those around
the stabilizer piece, which you will cut to fit later.

Stack up all the parts. Compare those that should
be the same: the two vees, the opposite tube sides,
the fork sides, the four cell the tube bulkheads.
Consider yourself an expert if the pieces come out
the same within ¹⁄₁₆″, and skillful if they're within
⅛″. If you have done your layout and checking
carefully, there should be no major errors in any
of the dimensions.

ASSEMBLING THE TUBE

Clamp the four tube sides together, edges aligned;
then plane or surform all four edges straight and
smooth. Unclamp the pieces, align the other edges
and fix them too. Smooth the saw-cut ends. Choose
the better side of each piece as the outside; then mark
the 5″-wide tube sides on the inside as A and B, the
5½″-wide tube bottom as C, and the tube top as D.

In the same way, plane the edges of the 5″x5″x¾″
plywood bulkhead and mirror cell pieces so their
sides match the widths of the 5″-wide sides of the
tube. (At this point, you may find that the nominal
5″ is actually 4¹⁵⁄₁₆″, but this small error will not
matter.) Cut 4½″-diameter holes in the two bulkhead
pieces with a compass saw or power saber saw; then
cut ¾″-radius corner notches, as shown in the plans,
in both cell pieces but in only one (whichever is less
well-made) of the baffles.

In bottom side C, drill two ⁵⁄₁₆″ holes 13″ and
19″ from the back end of the tube. Drill matching
⁵⁄₁₆″ holes 2″ and 8″ from the end of the plywood

tube-reinforcing piece.

In the top side D, use a hole saw, compass saw, or saber saw to cut a 1½″-diameter hole 43″ from the rear end of the side for the focuser. If you have received the focuser, drill mounting holes centered accurately around this hole.

Cut the ¾″ quarter-round molding into 43½″ lengths or pieces that total that length. Nail and glue them to the 5″-wide tube sides A and B with a ¾″ inset at each end. (This will leave ¾″ at both ends of the tube for insetting the reinforcing bulkheads, cut from ¾″ plywood.) Use ¾″ brads spaced about 2″ apart and a water-resistant glue such as Elmer's Carpenter's Glue. Set them aside to dry.

At last you're ready to put something together! Mark with a pencil sides A and B 21″ from the back end, run a bead of glue, then nail the corner-notched bulkhead to both with small brads. It really helps to have another person hold the sides while you drive the brads. Apply glue to the front bulkhead (the one *without* corner notches) and nail it in. Finally, just for a little extra strength, nail in the rear end cover—but *do not* glue it or drive the nails in all the way because you'll need to take this piece off later.

Check to see that this rather wobbly structure is close to square; then run glue along all the joint surfaces except the rear end cover. Place the tube bottom C on it, gently force side A and B to fit squarely against it, and then nail it in place with brads spaced every 2″. The third side will give the tube considerable strength. Turn the tube over; coat the tube-reinforcing piece with glue and place it on the inside of the tube bottom C. Insert the ¼″-20x½″ tee nuts through the two ⁵⁄₁₆″ holes at this time.

After the glue is dry, lightly nail the top side into place. Make sure that the focuser end goes at the front of the tube. Drill and countersink holes for 16 of the #8x¾″ flat-head wood screws; then drive the screws in. The tube is now a wooden box. If you want it to have rounded corners, this is the time to round them.

In preparation for finishing, remove the rear end cover. Fill any breaks in the grain or voids in the edge veneer of the plywood, and after the filler dries, sand the whole tube thoroughly and give it a coat of clear primer/sealer. Next, unscrew the wood screws hold-

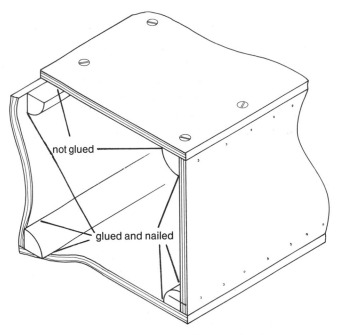

Cross-sectional view of the telescope tube. Note that the top side is attached with screws while the other parts are glued and nailed.

Smooth the sides, round off the corners, remove all rough spots and splinters.

Remove the fourth side in preparation for sealing and
painting.

ing it, remove the top side, and paint the inside of
both the tube and the top with primer/sealer.

INSTALLING THE OPTICS

By now, the optical parts should have arrived. Without getting fingerprints on the delicate front surface,
measure the diameter of the mirror as accurately as
you can. Also measure the focal length. Although it
should be 45″, or whatever the company you ordered
it from specified, the focal lengths of mirrors often
differ several inches from the advertised amount.
First, check to see if the focal length is written or
scratched onto the side or back of the mirror. If it
isn't, devise a temporary means of holding the mirror
vertical at one end of a long table. Punch a ¼″ hole
in an index card; tape the card to the front of a flashlight; then place the flashlight about 85″ away, shining
toward the mirror. Dim the lights and catch the image

of the hole on the other side of the file card. Slowly
move nearer and farther from the mirror until the
image of the hole is very sharp, then measure the
distance from the mirror to the card. *Half* of this distance
is the focal length.

After making these measurements, and while the
glue or sealers on the tube are drying, begin work on
the mirror cell assembly, which consists of the cell
and the support piece. Begin by carefully inscribing a
circle the exact diameter of the mirror on the mirror
cell. At three equally spaced points on the circle,
dent the wood with the point of a nail. Drill the ¼″
and ⁵⁄₁₆″ center holes. Chisel away one veneer of
plywood in the cell so the tee nut will be recessed,
and drill the holes for the tee nuts in the support
piece as shown in the plans. Insert all four tee nuts—
three in the support piece and one in the cell. Run in
the hex-head bolt and three round-head bolts. Turn the
hex-head bolt finger-tight. Check that the assembled
cell fits the rear end of the tube and will slide into the
tube for about 6″. If it's too large, plane, surform, or
sand it down. It should fit closely but not bind.

Before going further, prime and seal the cell, then
paint it and the inside of the tube black. A good
flat-black acrylic spray paint works well. Apply a light
first coat; allow it to dry; scour the parts vigorously
with a stiff bristle brush to remove lint, dirt, and loose
wood fibers; then apply another coat. Repeat this
until you have smooth, matte-black surfaces. Spray
paint is messy, so make sure you have adequate
ventilation and that the drifting spray doesn't blacken
your whole workshop. Let the paint dry thoroughly
before doing any more work on the tube.

Now for some math. You're ready to mount the
mirror cell, but first you must determine *where* to
mount it. The rule for a Newtonian telescope is
simple: *The distance from the mirror to the diagonal
plus the distance from the diagonal to the focal plane
must equal the focal length of the mirror* (see Chapter 2).
You already know the focal length, so you must first
find the diagonal-to-focal-plane distance. To do this,
measure the fully collapsed height of the focuser
(whether it's one you built or bought), then add 1″
because you want to allow room for some variation
between eyepieces. Add half the diameter of the tube,
2¾″, to this. The total is the distance from the focal

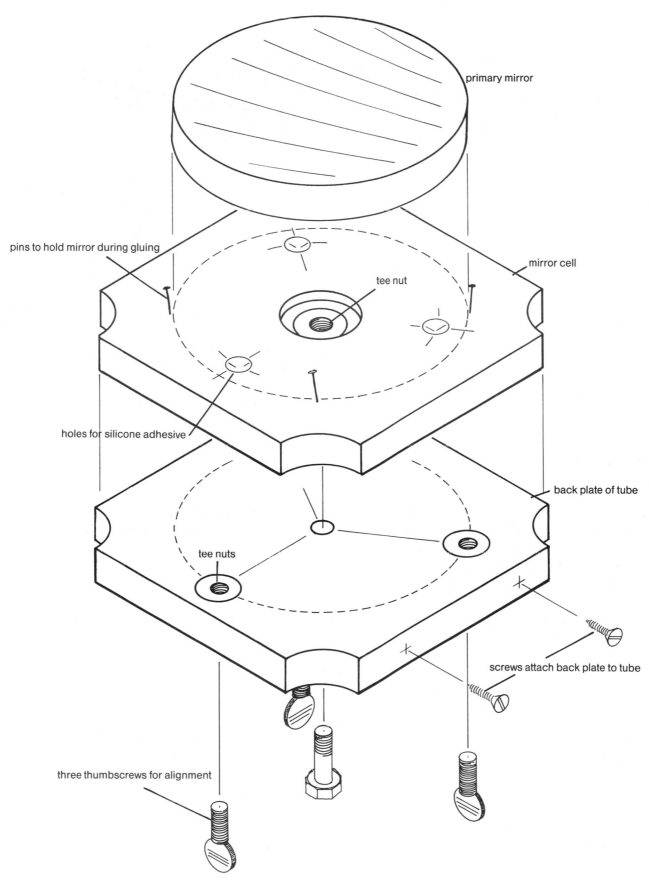

primary mirror

pins to hold mirror during gluing

tee nut

mirror cell

holes for silicone adhesive

back plate of tube

tee nuts

screws attach back plate to tube

three thumbscrews for alignment

Exploded view, showing how the parts of the mirror cell fit together.

plane to the diagonal; subtract this from the focal length of the mirror for the distance from the diagonal to the position of the front of the mirror. Since the center of the diagonal lies in line with the center of the focuser, measure back along the tube from the *center* of the focuser hole. Make a pencil mark there.

Now check for potential problems. If the mark is fewer than 4″ from the end of the tube, there won't be enough space for the mirror, cell, support, and rear cover. If this is the case, you can simply omit the rear cover or reduce it to a ring like the front reinforcing ring.

Next, drill three ¼″ holes ⅜″ deep about ⅜″ *inside* the indentations you made in the mirror cell. Blow all wood chips and debris out. Drive three brads about ⅛″ deep into the indents; then check to see that the mirror fits snugly between them. Place the cell flat on a table on newspapers and place three 6d finishing nails on the cell midway between pairs of brads. These nails act as spacers.

Squeeze a bead of silicone adhesive about ¾″-

Lowering the mirror onto adhesive blobs; small brads keep it centered in the cell.

diameter into and around each hole. Lower the mirror, *back side down*, into the cell between the brads. The adhesive should spread out into 1″-diameter blobs under the mirror. Some will escape out the side; push it back against the side of the mirror. Let it cure undisturbed for 24 hours, then gently remove the finishing nails. The mirror is now firmly attached to the cell with three soft, flexible silicone rubber pads the thickness of a nail. You may remove the brads if you wish since they have completed their function.

Next, cement the diagonal mirror to its support. Make a temporary gluing jig—even a stack of wood scraps about ¼″ shorter than necessary to support the mirror pad on the stalk facedown and flush will do. Place a pad of lens tissue (such as the paper the diagonal was originally wrapped in) on the jig; lower the diagonal on it facedown with the blunt, aluminized end of the mirror toward the stalk side of the support. Check to see that you can place the support pad flat against the back of the mirror; then squeeze a blob of silicone rubber adhesive onto the back of the diagonal, lower the support on it, and let it cure undisturbed for 24 hours.

(Alternatively, you may use Pliobond or another contact cement to glue the mirror to a piece of felt and the felt to the mirror cell. This gives a strong, flexible bond. Do not glue glass to metal or wood with epoxy or any other inflexible adhesive. The glass will be warped and the telescope will give poor images.)

Reassemble the cell and slide it into the tube until the front of the mirror lies opposite the pencil mark showing where it should be. Square the cell on so you can see your reflection in the mirror when you look down the front end of the tube. Tack the support piece in with several brads; drill and countersink holes; then anchor the cell with four #8x¾″ wood screws. Attach the side of the tube with its wood screws; then bolt the focuser and install the secondary mirror on its stalk. Install the rear cover plate with four #8x¾″ wood screws as you did the mirror cell.

ASSEMBLING THE MOUNT

The mount consists of two plywood assemblies, the base and the fork, joined by a "tee" made of pipe

The cell and primary mirror in place at the bottom end of the tube; silicone rubber adhesive holds the mirror in the cell.

The diagonal mirror is suspended in the center of the tube on a stalk, directly in line with the focuser tube.

fittings. It's best to assemble the base and pipe tee before completing the fork.

Start by planing or surforming the edges of the vees so the outside corners are rounded and smooth. Then fix up the edges of the large and small spreaders. Carefully measure the positions of the four U-bolt holes in the vees—note that they are offset 1″ from each other on the two vees—then drill them. After drilling, clamp the vees together and mark each inside surface with the location of the holes from the other vee. Separate them; then chisel recesses between the marks to the depth and width of the shank of the U bolt.

Now you're ready for a trial fit. Insert the U bolts loosely in each piece, hold the pieces together, push the 12″ pipe nipple through the U bolts until only 2″ protrude beyond the vees, and then snug the U-bolt nuts finger-tight. The pipe should rest firmly against both plywood vees; if it doesn't, the recesses must be enlarged. Slip the large spreader between the vees and tack it in with a few finishing nails. Tighten the

After drilling the U-bolt holes, chisel recesses for the opposite pair of U bolts.

The U bolts have been installed in the vees; note the position
of the chiseled recesses beside the U bolts.

Triangular spreader plates brace the vees.

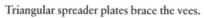

Four U bolts hold the pipe tee tightly in place.

fork sides together, and plane or surform them to the same shape and dimensions, then round the outer edges. Locate the pipe floor flanges on them. Drill a hole in each side for a tee nut at each hole in the floor flange: eight or ten holes in all, depending on whether the flanges have four or five holes. Finish these parts by gluing and nailing the counterweight arms (made up from two thicknesses of ¼″ plywood) on and letting them dry.

Drill the remaining holes shown on the plans. The two on the mounting plate, ⁵⁄₁₆″ in diameter and 6″ apart, must line up with the holes on bottom side of the tube.

Next, assemble the pipe tee. Thread the two sort nipples onto the tee fitting and the floor flanges onto the nipples, and then the entire assembly onto the 12″ nipple protruding from the base of the mount. They should be run on tight enough that they turn with moderate difficulty but not so tight they'll jam if turned less than a full turn farther.

Bolt the fork sides onto the flanges with ¼-20 flat-

Attach the fork sides with wood screws, but do not glue this joint—you may need to adjust it later.

U bolts until they grip the pipe tightly. Measure, mark, and fit the stabilizer between the lower ends of the vees. You must bevel three edges of the stabilizer: a 45° angle where it contacts the spreader, and approximately 112° angles against the vees. Use a rasp, plane, or surform. When the stabilizer fits all around, tack it and the small spreader in place with finishing nails.

Drill and countersink holes for 18 #8x2″ wood screws, nine on each side: two for the small spreader, three for the stabilizer, and four for the large spreader. Loosen the U bolts, knock out the finishing nails, apply wood glue to the joints, then reassemble the mount with the wood screws. Tighten the nuts on the U bolts until the wood around them crushes slightly. You'll hear it crunch!

Now you're ready to work on the fork. Clamp the

Close-up of the completed polar
and declination axes; note also
the thumbscrews that attach the
tube to the mount.

head bolts and tee nuts, then measure the distance
between the flanges. Trim the mounting plate to this
width and slip it between the sides. Tack it in place
with finishing nails; drill and countersink holes for the
remaining six #8x2″ wood screws and run them in
tight. Do not glue these joints. You may need to
readjust them later as the pipe threads wear in.

You will need to use some ingenuity to find two
counterweights to balance the telescope. Each should
weigh about two pounds. Scrap metal is good, but
boxes or cans filled with small nails or lead shot are
best. Don't seal them tight—they should be adjustable.
Bolt them to the counterweight arms.

Now for the first trial assembly. Attach the tube to
the mount, using a thumbscrew, hex nut, and fender
washer through the fork and into the tee-nutted holes
in the bottom of the tube. Push the tube gently: It
should move smoothly and without binding. You can
adjust this by tightening or loosening the flanges. Is
it balanced? If it is not, find heavier or lighter counterweights. If the front of the tube is too heavy, you
may have to add weight to the inside of the rear cover

plate. Lead shot is excellent. Check to see that everything fits. If it doesn't, this is the best time to correct it.

FINISHING AND PAINTING

Your telescope is now almost usable. At this point
you may want to try using it for several nights. If so,
skip ahead to the next section on collimating the optics,
and then you're ready to go. However, for appearance's
sake as well as better weather resistance, it's a good
idea to prime/seal and paint all exposed wood. Fill
the edge voids with a thick mixture of wood glue and
sawdust, and let it dry. Sand it flush, fill it with wood
filler, and then sand it again, along with the rest of the
surface. When it's smooth, vacuum it clean, remove
residual dust with a tack cloth, then apply a generous
coat of primer/sealer. Be sure the edges of the plywood
are sealed; apply two coats if necessary. Sand again
lightly, then paint with polyurethane varnish or any
color of your choice. Polyurethane floor enamel is the
best telescope paint stocked in hardware stores. It is
both tough and waterproof. Choose a relatively light

Attach the finder to a convenient spot on the fourth side of the tube.

The moon will probably be your first target as you begin exploring space with your
new telescope! This photo was taken through the 4″ reflector.

color so you can see the telescope in the dark.

You may also want to paint the diagonal support and the rim of the diagonal with flat black paint even though this is not strictly necessary. Apply the paint with a small brush. Don't spray because if you get paint on the aluminum surface, you can't get it off.

COLLIMATION

The last operation you must perform before you can actually *use* your telescope is collimation. What you're trying to accomplish in collimation is to direct the optical axis of the mirror through the exact center of the tube, intersect the exact center of the diagonal, through the center of the focuser, and through the optical axis of an eyepiece. Luckily, it's not as tough as it sounds; it's just a matter of following a set of collimation steps.

Step one: Slide the diagonal support stalk back and forth until the diagonal lies exactly in the center of the tube. Check this with a short ruler held inside the tube.

Step two: Rotate the diagonal stalk so that by looking into the focuser (with no eyepiece in it), you see the primary mirror reflected in the diagonal. Center the image of the primary in "rotation." Don't adjust the "tilt" yet.

Step three: With one eye open several feet behind the focuser tube, stare straight down it. The outline of the diagonal should be concentric with the front and back ends of the focuser tube. Gently bend the diagonal support stalk (use fingers for small adjustments, pliers for bigger adjustments) until it is.

Step four: Check to see that the adjustments you made in steps one and two still hold, and recenter and realign the stalk if necessary.

Step five: Make an alignment tool, a small cardboard or plastic plug 1¼″ in diameter that fits into the focuser tube. In its *exact* center, drill a hole ¹⁄₁₆″ in diameter. The alignment tool assures that your eye is exactly centered when you look down the focuser tube. Insert the plug into the focuser tube.

Step six: Only one error in the diagonal adjustment should remain, and that is the "tilt." Gently bend the stalk until the image of the primary is exactly centered in the diagonal mirror.

Step seven: Recheck the adjustments you originally made in steps one and three. If they are still correct, you will probably never need to repeat them.

Step eight: Remove the rear cover plate.

Step nine: With the alignment tool in place, look at the image reflected in the primary mirror. Your goal is to exactly center the *reflection* of the secondary mirror in the primary. Loosen the central "pull" hexhead bolt a little, then turn the three push bolts on the back of the cell to accomplish this. Retighten the central pull bolts.

Step ten: Replace the rear cover plate.

Even if you give it pretty rough treatment, chances are that the telescope will remain aligned for several years before slow dimensional changes necessitate repeating steps eight, nine, and ten. Even then, you'll probably find the total adjustment will amount to ¹⁄₁₀ of a turn on one bolt and take only a minute or two.

A FEW HINTS ON USING YOUR NEW TELESCOPE

It's done! Now you can see how your telescope earned its nickname, "the squatting dog." You'll soon know the thrill of carrying the tube and mount out the back door, into the yard, and bolting them together. What next? If you've taken to heart the advice at the beginning of this chapter, you've learned the bright constellations and stars, and have a star map that shows some of the deep-sky wonders. You won't be lost under the stars.

Place the mount on the ground so the polar axis (i.e., the 12″ pipe) points to the north. Pop the eyepiece in. The moon should look bright and be crisply defined. Stars should be tiny pinpoints; Jupiter's satellites should be readily apparent; and Saturn should appear as a tiny and jewellike planet set in its rings. Venus should show its phases; star clusters and nebulae should appear sparkling and delicate (depending to some extent on how good your sky is); double stars should be resolved right to the theoretical limit.

If you have trouble in finding sky objects by sighting along the tube, add a peepsight finder or finder telescope. Wait awhile before ordering more eyepieces. See all the sights you can at low magnification before going for "more power."

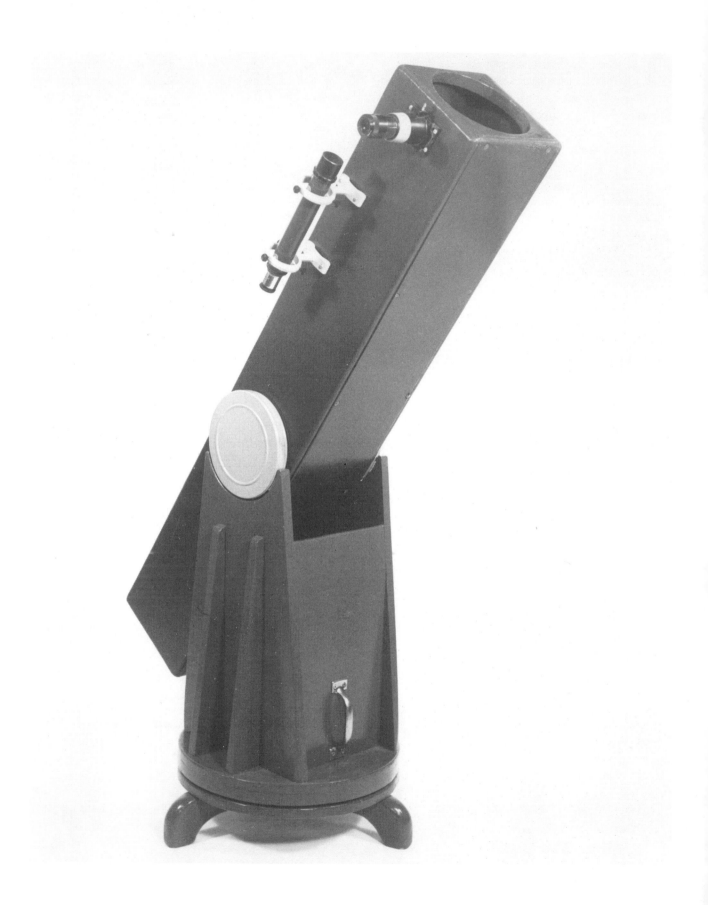

This Dobsonian reflector is a rugged, compact, portable, and easy-to-use instrument capable of showing all the basic wonders of the universe. The telescope moves on buttery-smooth Teflon plastic and Formica bearings. The tube and mounting are made of plywood.

Building the instrument is a straightforward job; it involves only simple carpentry. The finished product stands up to use by enthusiastic teenagers, car-camping trips, lots of star parties, or years and years of backyard enjoyment. The tube is a plywood box weighing just 18 pounds. The mount weighs 20 pounds and stands just over 24″ high. One person can easily carry both of them out for a night of observing.

PLANNING

The very first step in building this telescope is to order (or start grinding) a 6″ f/8 mirror. The plans assume that its focal length will lie between 44″ and 52″, well within the tolerance of commercial f/8 mirrors, so as soon as you order the mirror, you can begin construction with reasonable confidence that the mirror will be suitable. It is well within a novice mirror-maker's ability to finish a mirror to a focal length within a +/- 2″ range, although the focal length will need to be tested frequently during fine grinding (see Chapter 10).

If the focal length is longer than 52″, the focuser will be too close to the end of the tube, and you will need to make the tube longer and increase the height of the rocker box to accommodate it. If the focal length is under 44″, you may shorten the tube if you wish to, but the extra tube length won't hurt anything and is, in fact, quite useful because it acts as a light shield.

The tube of the telescope is made of ½″ plywood and is intentionally that thick. Although thinner plywood would be strong enough, the resulting tube would be so light that it would turn too easily on its bearings. The heavier tube gives better stability and will stand up to rougher handling than a thinner tube; and at 18 pounds, it is really not very heavy.

There are two minor variants on the mount. If you plan to use the telescope mostly on grass or soft surfaces, build the version with three short legs that

Building a 6″ f/8 Dobsonian Reflector

5

will raise the base a few inches off the ground. If you plan to use it mostly on bare ground or pavement, omit the legs and simply use feet.

GATHERING SPECIALIZED MATERIALS

Besides the primary mirror, you'll need a diagonal mirror (1.25″ minor axis), one or more eyepieces, and an eyepiece holder, which you can purchase from one of several suppliers listed in Appendix B or make yourself (see Chapter 9). You'll also need two round objects to serve as side bearings. These must be between 6″ and 8″ in diameter, about ½″ to 1″ thick, have a smooth, hard rim, and be accurately round. Old 45-rpm record-player turntables are excellent, and you can salvage them secondhand or at rummage sales. You can also make the bearings from a ¾″-plywood disk by cementing a strip of aluminum, stainless steel, or Formica to the outside. The handiest ready-made bearings, though, are metal 16mm film cans, available for about $2.50 each at any well-stocked camera store.

You'll also need seven small pieces of Teflon plastic ⅛″ thick for the bearings. Three of them should be

The 6″ Dobsonian telescope is extremely portable: One person can easily carry it outdoors for an evening of sky-gazing.

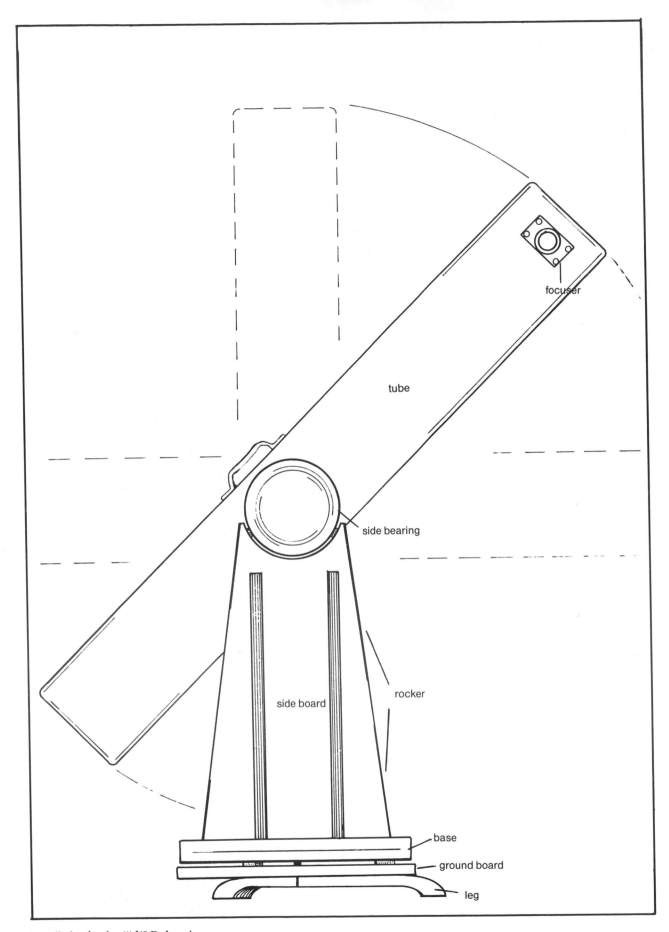

focuser

tube

side bearing

rocker

side board

base

ground board

leg

Overall plan for the 6″ f/8 Dobsonian.

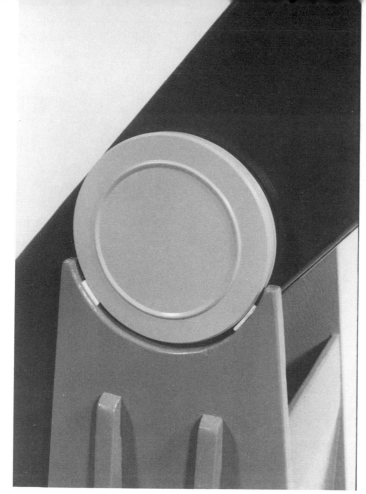

The telescope moves with buttery smoothness on inexpensive movie film cans.

Wood, optics, and hardware—elements of a simple, yet effective, telescope.

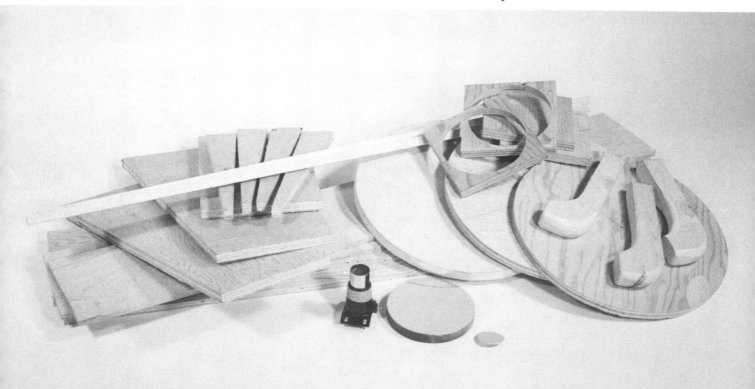

roughly 2″x2″, and four should be ½″x1″. The biggest problem is that this is such a small quantity of material that it's hardly worth anyone's while to sell it. See Appendix A for hints on finding a local supplier.

CUTTING THE PARTS

The wood parts for this telescope can be cut from two half sheets (4′x4′) of exterior-grade plywood, one ½″ thick and the other ¾″. If you buy full sheets, you'll have plenty of wood left for other projects. You will also need a sink cutout (usually Formica bonded to ¾″ particle board) at least 16″ in diameter, 16′ of ¾″ quarter-round molding, and 3′ of 2″x4″ lumber for the legs.

Begin by cutting the panels for the tube. Use whatever saw is best for you. A handsaw will do a fine job if you take care to make the cuts straight. For the sake of appearance, the grain of the wood should run in the long dimension of the tube. Cut the two side panels 8″x48″; cut the top and bottom panels the same length but 7″ wide. Clean up the edges after cutting them.

Next, saw the rocker-box side panels from ¾″ plywood. Cut a piece of 13″x42″, with the grain running in the long dimension. Then carefully measure and mark the centers of both dimensions and make the angle cuts shown. Mark the slender triangular reinforcing pieces 5″ from their thin ends. The small triangles remaining above this are too thin to have any strength and so are scrap. Locate the centers of each circular cutout 1″ beyond the center line and swing a radius ⅛″ greater than the radius of the side bearings. For example, for the 7¼″-diameter metal film-can side bearings in the prototype, the radius should be 3¾″. When you cut the parts, take care not to splinter the triangular reinforcers. Here a handsaw is helpful because you can cut slowly. A saber saw is best for the circular cutouts, but a hand-held keyhole saw or a coping saw will do a perfectly adequate job. After cutting them, clamp the matching side panels together and plane them to the same dimensions.

Three 16″-diameter circles follow, two cut from ¾″ plywood and one from the sink cutout. If you don't have a compass large enough to mark them, drill a ¼″ hole in a piece of scrap, and 8″ from it, drill a ¹⁄₁₆″ hole. Force a pencil through the ¼″ hole and a nail through the small hole, then use this makeshift compass to draw the circles. Cut these with a saber saw or a keyhole saw. Cut carefully: The telescope will work perfectly well if the circles are not exactly round, but it will look much better if they are.

Cut the four 7″x7″ bulkheads and the 8½″x16″ front board from ¾″ plywood. These are all straight cuts. From two of the bulkheads, cut 6½″-diameter circles. One of the remaining bulkheads, the slightly larger of the two, will become the end of the tube; the other will be the mirror cell. Drill three ⁵⁄₁₆″ holes in the tube end as shown on the plans and install ¼″ tee nuts in them for three push bolts. Drill a ⁵⁄₁₆″-diameter hole in the center of the cell and a ¼″ hole in the tube end; then install a ¼″ tee nut for the pull bolt. Countersink all four holes 1″ diameter by ¼″ deep in the tube end piece.

Finally, if you're building the "grass-and-soft-ground" version of the mounting, cut off three 10″ lengths of 2″x4″. Saw out pieces 1½″x7″ to approximate their shape, and use a large rasp or surform plane to rough out the curves of the legs. Be sure to leave the inner 8″ of the top surface flat since this is the gluing surface. The feet for the "bare-ground-and-pavement" version are three 2½″-diameter circles cut from the mirror cell with a hole saw.

ASSEMBLING THE TUBE

Begin the tube by gluing a strip of quarter-round molding exactly ½″ inside each edge of the 8″-wide tube side panels. Each quarter-round must end ¾″ inside each end, for a total length of 46½″. Place the 7″-wide top panel on the workbench with the side panels on either side of it; tack them together lightly to make a three-sided box, then check the fit of the bulkheads. Place the better of the two with the opening at the front of the box with the other in the middle. Pick the slightly larger of the uncut bulkheads as the rear bulkhead; the other will become the mirror cell. Tack these pieces in place with brads, and check to see that the assembly is square. Open the joints again,

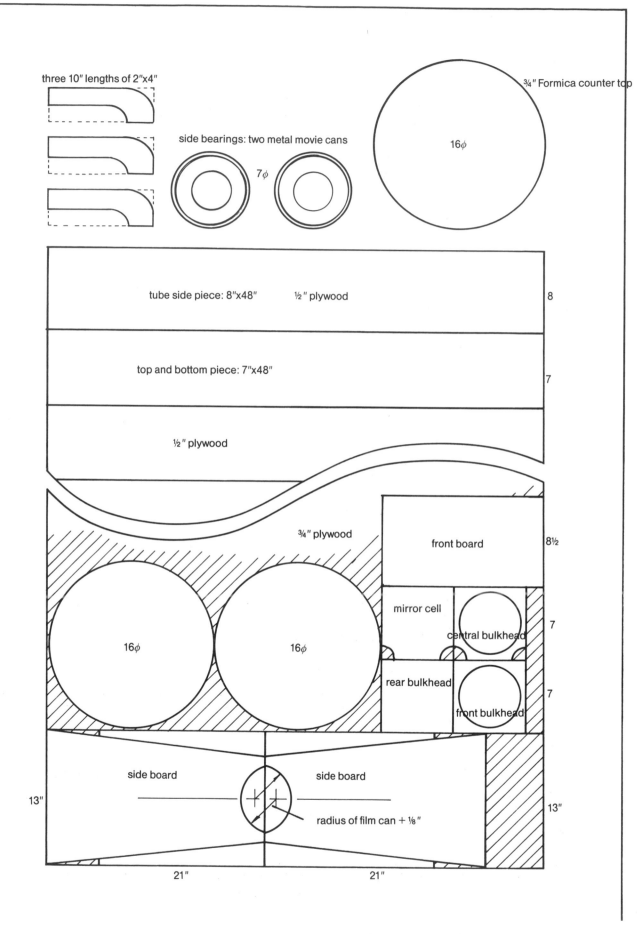

three 10" lengths of 2"x4"

side bearings: two metal movie cans

7φ

¾" Formica counter top

16φ

tube side piece: 8"x48"   ½" plywood

8

top and bottom piece: 7"x48"

7

½" plywood

¾" plywood

front board

8½

mirror cell

central bulkhead

7

rear bulkhead

front bulkhead

7

16φ

16φ

side board

side board

radius of film can + ⅛"

13"

13"

21"

21"

The parts for the 6″ f/8 Dobsonian can be cut from two half-panels of plywood and a sink cutout.

apply glue to both surfaces, then close them and clamp the tube until the glue sets. (This job is finished so quickly that it's hard to believe!)

When the glue has dried, use a rasp or surform to bevel the sides of the mirror cell so it will fit into the tube with only a slight clearance on all sides. With a hole saw or a saber saw, make the large ventilation openings in the cell (these can be the feet), and drill three ¾″-diameter holes, equally spaced, on a radius ⅛″ greater than half the diameter of the mirror (approximately 3⅛″ radius, but check your mirror).

Cut three 1¾″ lengths of ¾″ dowel; then carefully saw halfway through them 1″ from the end and saw out half of the dowel. This leaves ¾″ of intact ¾″ dowel. Drill a ³⁄₁₆″ hole ¼″ from the end of the

dowel, then countersink it from the flat side. Sand the dowels and round all the corners. If there are any splits or defects in a dowel, make a replacement for it. (Note: The photographs show a longer dowel with two holes in it. The extra length and second hole are not necessary.)

Glue each dowel into the mirror cell. If the dowel is slightly undersized in the hole, build it up with a layer of paper; if oversized, sand it down to a tight fit. Countersink a ¼″-deep hole ½″ inward from each dowel: a cup for adhesive. When the glue dries, sand or surform the dowels flush. Where the push bolts will press against the cell, attach a small mending plate to protect the wood. Finally, coat the cell with primer/sealer, let it dry, then paint it flat black.

The telescope tube, from above, showing the primary mirror in its cell, the side bearings, and the diagonal and focuser.

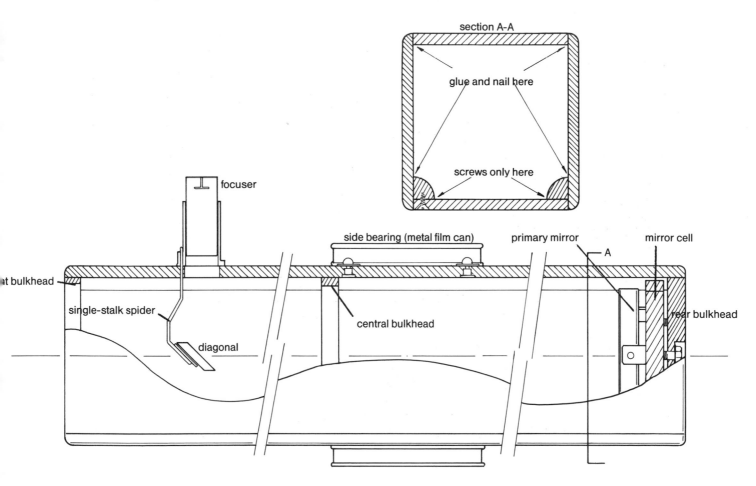

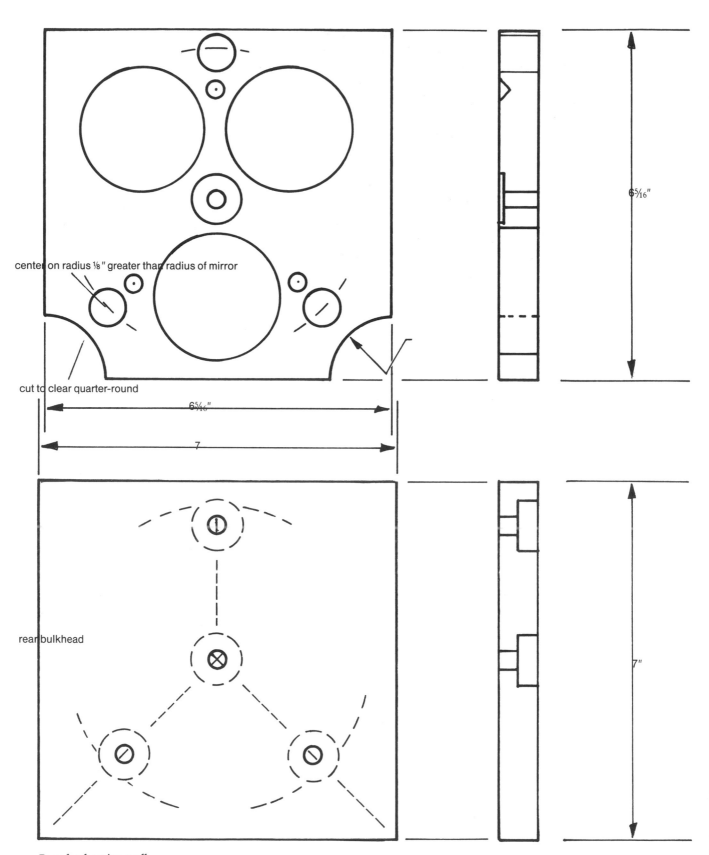

center on radius ⅛″ greater than radius of mirror

cut to clear quarter-round

6⁵⁄₁₆″

7

6⁵⁄₁₆″

rear bulkhead

7″

Parts for the mirror cell.

One of the three mirror-support dowels has been glued into the mirror cell.

The mirror cell installed in the tube.

Meanwhile, slip the bottom side into the tube assembly, tack it in with small finishing nails, then drill a set of countersunk pilot holes for #8x1¼″ flat-head wood screws. With all four sides in place, round off the corners of the tube and sand the entire outside smooth. Remove the bottom of the tube again. Prime the inside and outside of the tube, let it dry, fill any voids with cellulose wood filler, then prime it again.

The next essential step is to determine the focuser's position. Measure the height of the fully collapsed focuser (typically 3″ for friction focusers); add 1″ to allow for variation between eyepieces; then add half the diameter of the tube, or 4″. Subtract this figure (about 9″) from the focal length of the mirror. This is the distance from the mirror surface to the center of the focuser hole. Add the thickness of the mirror to that, then measure off that distance along the tube from the mirror cell. Mark the spot, then cut an appropriate-sized hole in one of the sides of the tube with a hole saw or a spade bit and drill pilot holes for mounting the focuser on the side of the tube.

At this stage, determine where the tube will balance.

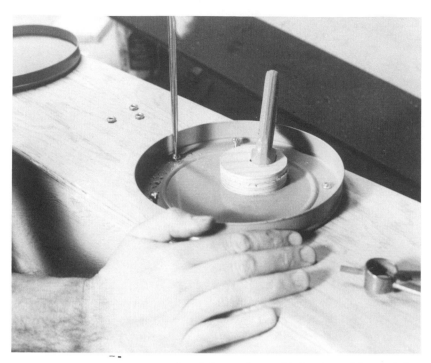

Attach the movie-can side bearings with stove bolts.

Long wood screws hold the base to the sides of the rocker box.

Place the cell, mirror, focuser, diagonal, finder, hardware, and bottom side in the tube opposite its final location. Place a dowel under the tube and roll the tube back and forth until you locate the "teeter point" where it balances. Mark this spot.

Remove the accessories and unscrew the bottom of the tube again; then clamp the side bearings to the outside of the tube. Depending on the material they're made of, attach them appropriately. For metal movie film cans, drill six equally spaced holes through one side of the film can and into the tube, then bolt the can to the tube. After you're sure the bearing is in the correct spot, glue on the other side of the can. For the time being, however, remove the bearings. Paint the inside of the tube and bottom side with two coats of flat black paint and the outside with two coats of polyurethane floor enamel or a durable coating of your choice.

ASSEMBLING THE MOUNTING

The mounting now consists of a pile of wood—big circles and side boards, slender triangles, and thick legs. We will assemble it first, checking for fit, with finishing nails; then we'll come back with glue and wood screws.

Bevel the edge of the base with a plane or surform tool.

The pivot hole in the base should be perpendicular to the surface.

The mounting pivots around the center bolt in the bottom bearing.

Begin by nailing the side boards, good side out, to the front board. It's crucial to get the side boards symmetrical and the front board upright. Turn these upside down, then nail one of the plywood circles to them. Double-check to see that the rocker is square and that the separation between the side boards is the same everywhere (if it's not, adjust it, then reassemble the rocker). Pry the joints open again, squeeze in glue, nail them shut, drill pilot holes, and screw the rocker together with #8x2″ wood screws. When the glue sets, glue and screw two of the slender triangular reinforcers to the outside of each side board.

When the glue is dry, apply glue to the bottom of the circle and the particle-board side of the Formica circle, clamp them to the rocker and again allow the assembly to dry.

Shape the feet (or legs) to their final form with a rasp or surform, then sand them smooth. Nail them to the second circle, equally spaced; check and fix the fit, then glue and screw the base together. Let the glue dry.

Drill a ¼″-diameter pivot hole exactly perpendicular through the center of the rocker-box bottom and the base circle. If you have a boring jig, use it; otherwise check with a try square to see that the drill is straight, and do the best you can to keep it straight while drilling. Enlarge the first 1″ of the hole in the base

Fill any edge voids with filler,
then sand smooth.

The three major parts of the
telescope—the tube, the rocker,
and the base—assembled and
almost ready to paint.

A coat of paint will prevent moisture from reaching the wood.

Recess the heads of the brads holding the Teflon pads by half the thickness of the Teflon. (above)

The focuser and diagonal holder may vary from 2″ to 8″ from the front end of the tube; shown is the minimum allowable distance. (below)

from the bottom side and install a ¼″ tee nut in it.

Fill edge voids and surface holes with cellulose filler; when it has dried, sand the rocker box and base smooth and apply two coats of primer/sealer to the wood. Do not get primer/sealer on the Formica bearing surface of the rocker. Finish the rocker box and base with two coats of polyurethane floor paint and let them dry thoroughly—several days in hot sunshine—before proceeding.

All that remains is to nail the Teflon pads to the side boards and base. Use brads about ¾″ long. Gently drive the brad down to the surface with a light hammer; then use a nail set or the tip of a large nail to bury the head of the brad about halfway through the Teflon pad. Assemble the mount with a 3½″ hex-head bolt through the pivot hole; then determine where it balances and add a large carrying handle.

Step 1: Squeeze a blob of silicone adhesive into each of the three countersunk holes in the mirror cell.

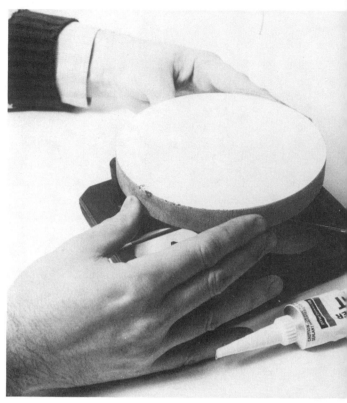

Step 2: Lower the mirror onto three adhesive blobs; 16d nails act as spacers.

INSTALLING THE MIRRORS

You are now ready for final assembly of your telescope if the optics are finished or if they were bought ready-made. If you are making your own mirror, complete the tube and mount to this step for testing the mirror on stars. Determining the focuser position will have been difficult because you knew the focal length of the mirror only approximately when you drilled the hole, but if you guessed reasonably well, you can now observe bright stars to check the quality of your mirror as you "figure" it. If necessary, plug the first focuser hole and drill a new one.

Begin final assembly by gluing the diagonal mirror to the diagonal holder with three blobs of silicone rubber adhesive (the kind sold as aquarium cement sticks to glass especially well). For a stalk-type of diagonal holder, prepare a gluing jig: Stack plywood scrap until the diagonal-holder surface rests flush against the top piece, then remove ⅜″ from the height of the stack. Place the diagonal facedown on a small square of optical tissue atop the pile, squeeze three blobs of adhesive ⅜″ in diameter on the back, then gently rest the holder on the blobs so they squeeze out ⅛″ thick. Let the adhesive set for 24 hours before disturbing the mirror.

Bolt the focuser to the side of the tube, then rotate, slide, and bend the diagonal stalk gently until it satisfies the following conditions:

1) The mirror is centrally located in the tube.
2) The mirror center is in line with the center of the focuser tube.
3) The bottom of the tube appears centered in the diagonal mirror.

If you are using a commercial diagonal holder and spider, follow the diagonal alignment directions for the polar plate reflector (Chapter 6).

To test a mirror, simply wrap a few turns of masking

Step 3: Squeeze silicone adhesive through the holes in the mirror holders.

To make collimation adjustments, use a socket screwdriver to turn the hex-head bolts that support the mirror cell.

tape around the mirror and over the mirror holders sticking out of the cell, then place the cell in the tube. Put in the push-pull bolts and align the mirror so that when you look into the front end of the tube, the front of the tube and your own reflection are centered in the polished face of the glass.

For permanent installation of the mirror, purchase a tube of silicone adhesive (bathtub caulk) or aquarium cement. Place the cell in a clean work area and lay three 16d common nails on the cell to serve as spacers. Squeeze three ¾″-diameter blobs of adhesive into the countersunk holes, then gently lower the mirror into the cell. It will contact the blobs and flatten them to the diameter of the nails. Carefully center the mirror with your fingertips. Insert the nozzle of the adhesive tube into the holes in the holders and squeeze a bead of glue through. Continue squeezing until each blob is about ¾″ in diameter, but don't allow adhesive to spread onto the surface of the mirror. Allow the glue 24 hours to cure before disturbing the cell.

Then slip the cell into the tube, run the adjusting bolts in and tighten them until the reflection of the open tube end is centered in the mirror. Screw on the side of the tube.

Now look into the focuser tube. You should see the reflection of the mirror centered in the diagonal mirror. If it isn't, adjust the diagonal until it is. This is a frustrating operation because the adjustments are not independent—changing one changes the others— but stick with it. Finish by adjusting the primary mirror to reflect the diagonal from its center. Later you can improve this preliminary eyeball alignment (see Chapter 11), but it'll be good enough now to show you some fine celestial sights.

Bolt the finder to the telescope tube and the rocker to the base. Attach handles and any other miscellaneous hardware. Carry the telescope and mount outside, set the mount down and set the tube onto the bearings. You're now ready to start observing!

On a dark, clear night in early fall, the star clusters in Auriga could be the targets for your newly completed telescope.

Our second 6″ is a Newtonian mounted on a polar-plate German equatorial, a modern version of fine instruments made in the late nineteenth century. The telescope is sturdy enough for high-magnification observing, yet it moves with a light touch. The mount can be adjusted for precise polar alignment, and if your add the optional slow motion, it tracks the sky as you turn a control knob.

The optical tube is a modernized classic: a thin tube stiffened by external rings and rotating in a cradle. The mirror is held in its cell with silicone rubber rather than by metal clips, and the cell is mounted with push-pull bolts. Once you've aligned it, it'll stay aligned.

The mount consists of two major parts: the pier and the equatorial head. The pier is a hollow wooden box attached to a heavy, triangular base that supports it and the telescope solidly. You should adjust the height of the pier if you plan to mount a telescope significantly longer or shorter than the standard 6″ f/8 reflector.

The equatorial head sits atop the pier. The polar axis consists of two Formica-surfaced disks separated by a sheet of Teflon plastic and held together by a stiff spring. It is aligned on the pole with push-pull bolts. The declination axis is made from a length of steel pipe running in wooden V blocks and Teflon.

Besides the locally available materials and hardware listed below, you will need to make or purchase a 6″ f/8 objective mirror, a diagonal mirror (1¼″ minor axis), a focuser for 1¼″ eyepieces, a spider to hold the diagonal mirror, a finder, and one or more eyepieces. These are described in Chapter 9.

PLANNING CONSTRUCTION

How you go about building this telescope depends largely on you. You might build all the major parts more or less simultaneously so that at the end, the product of your labors will emerge quite suddenly. This is actually very efficient since many steps require letting glue set or paint dry for several days before going on. If you are grinding the mirror, finish the mount first (perhaps alternating carpentry with rough and fine grinding), then complete the tube just enough to allow you to test the mirror on stars during the final figuring.

Building a 6″ f/8 Equatorial Reflector

6

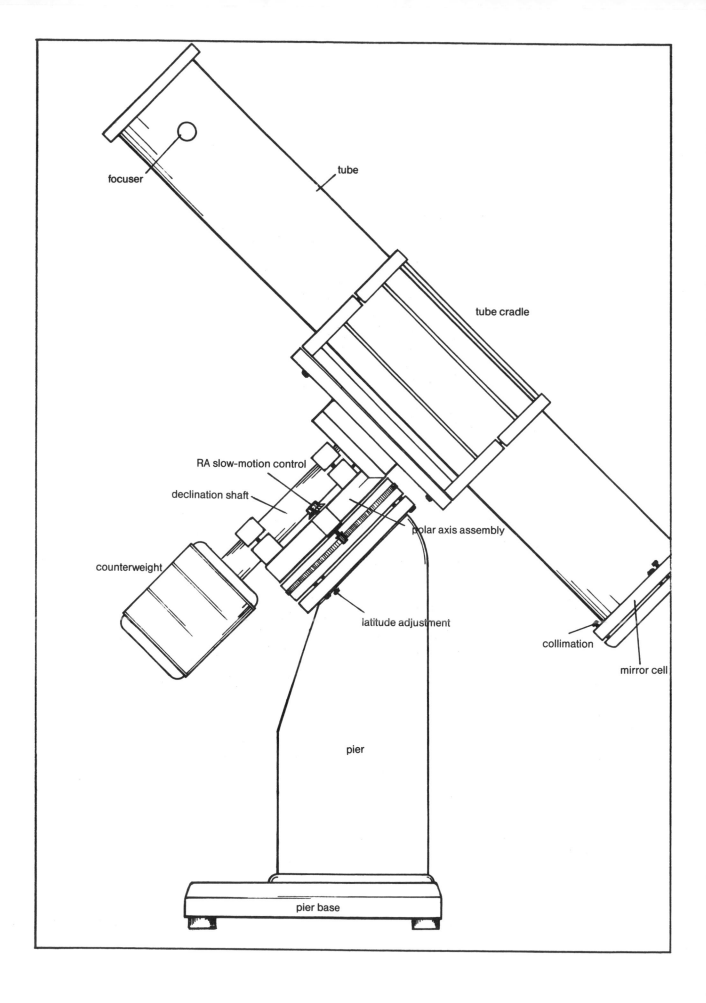

focuser

tube

tube cradle

RA slow-motion control

declination shaft

polar axis assembly

counterweight

latitude adjustment

collimation

mirror cell

pier

pier base

All the parts for the telescope and mount can be cut from a single 4′x8′ sheet of ¾″ G2S plywood and a Formica-covered ¾″ particle-board sink cutout. Begin by laying out the base and pier. The side pieces of the pier are mirror images of each other. The angle between the back of the pier and its top will be your latitude; i.e., for latitude 40°, the angle between the pier back and the top is 40°. Cut the three front and back panels, the bottom panel, the interior panel, and the top panel to exactly the same width.

The pier base parts have rounded corners. With a saber saw cut these round. With a table saw, cut triangles and remove the corners later. The pier is topped by a 10″-diameter pier plate A, grouped on the layout plan with the equatorial head parts. Cut it out with the circles for the equatorial head.

The equatorial head consists of three plywood circles, four Formica sink-cutout circles, and one rectangular plywood piece. Mark the parts in dark pencil (be sure to "save" the center by making a dent in the wood), then cut them out with a saber saw. Try to make the cuts as smooth and precisely round as possible.

The counterweight requires three plywood circles,

Just a pile of wood! The equatorial mount before assembly.

W1, W2, and W3, all roughly 6″ in diameter. The dimensions depend on the diameter of the cylinder chosen for the counterweight. If you use a 6″-diameter, three-pound coffee can, W1 and W2 are 5⅞″ in diameter, and W3 is 6⅛″ in diameter. W1 and W2 fit inside the can rim; W3 overhangs it.

In addition to the wood, gather the following:

- 2′ of 1½″-square clear pine molding for the declination bearings
- 3′ of ½″ quarter-round molding for decoration on the base of the pier
- a 12″ length of 1½″ galvanized pipe, threaded both ends
- two 1½″ galvanized floor-flange pipe fittings
- two square feet of 1/32″ thick sheet of Teflon plastic
- a ½″x6″ hex-head bolt plus matching non-slip nut and two fender washers
- a stiff spring 1½″ long and 1″ in diameter (a valve-lifter spring is suitable)
- four ¼″-20x4″ carriage bolts plus nuts and washers
- three 1½″ and three 1⅜″ hex-head bolts
- six ⅜″ tee nuts
- a ¼″ threaded rod
- lots of ¼″-20 nuts, washers, and fender washers
- a dozen ¼″-20 tee nuts
- a variety of miscellaneous hardware

Be prepared to make numerous trips to the hardware store—there's a fair amount of "cut and try" work in this design.

If you want a right-ascension, slow motion control, obtain a large knob, a 6″ length of ⅛″x1″ flat steel, a few inches of ¼″ steel rod, a ⅜″- or ½″-wide rubber timing belt with 5 teeth per inch, a 12-tooth plastic timing gear to match the belt, and a bottle of contact cement.

ASSEMBLING AND FINISHING THE PIER

Clamp the pier sides together and finish them to the same dimensions. Clamp the five panels together

and plane or surform them all to the same width. (If you are using hand tools, switch the order of the stack from time to time so the pieces don't get rounded edges.)

Experimentally assemble the panels on one of the side pieces, adjusting any minor fitting problems with surform or plane. Drill the holes in the bottom, interior, and top pieces. Reassemble the panels on one side piece, check that all the angles are square, then coat their upper sides with wood glue, lower the other side into place, and tack it together with finishing nails. Turn the assembly over, then glue and tack on the other side. Drill pilot holes and finish the assembly with a dozen #8x1½″ flat-head wood screws per side.

While the pier dries, glue the three basic triangles into a single massive base slab. Spread glue across both surfaces lightly, bring them together, and clamp them tightly. It is unnecessary to reinforce such large areas of contact with screws or nails.

When the glue has set, plane, surform, and sand the edges smooth; then fill any voids in the plywood with wood filler. To round the contours of the base, use a large wood rasp. Cut a cardboard template to the edge profile, then rasp away until the base conforms all the way around. Finish with a surform or coarse sandpaper; then fill any voids and sand the edges smooth after the filler sets. Drill holes for the tie-down bolts and for the foot bolts; then cut 2½″-diameter plywood disks for the feet and glue them in place. Apply a coat of primer/sealer to every surface you can still reach and allow it to dry thoroughly.

Center the pier on the base and draw its outline in pencil. Drill pilot holes from this outline into the base from the top, then turn the base over and drill ½″ holes ¾″ deep. Place the pier on the base, install tie-down bolts through the interior panel, and tighten them. Drill pilot holes from the ½″ holes into the pier. Loosen the tie-down bolts, put glue on the base of the pier, then tighten the tie-down bolts and run #10x2″ flat-head wood screws through the base and into the pier. Cut and fit lengths of ½″ quarter-round molding around the base of the pier, then glue and secure them with brads.

At this stage, the bulk of the work on the pier is complete. Although disk A is really part of the pier, it should be constructed as part of the equatorial head and added to the pier later. Finish the pier as you

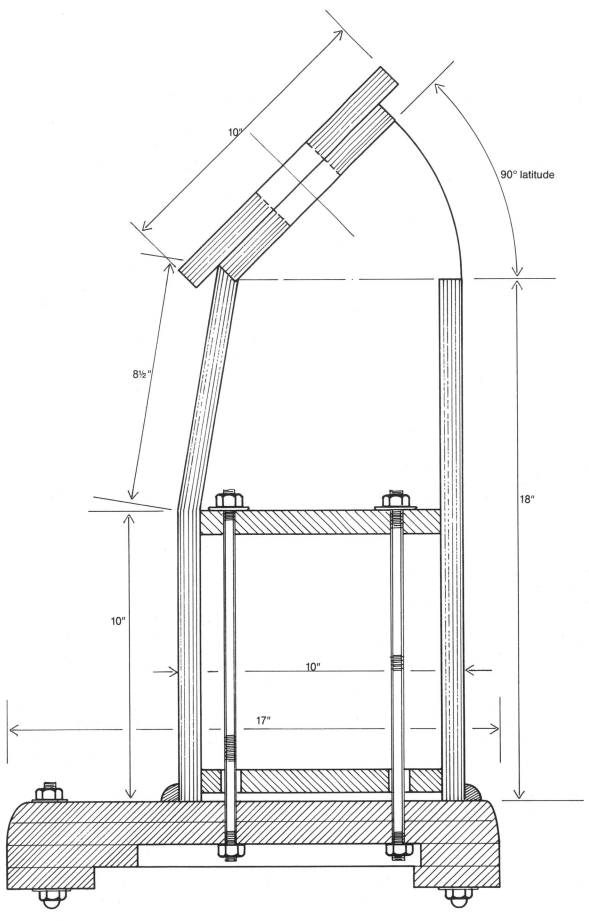

10″

90° latitude

8½″

18″

10″

10″

17″

Cross section of the equatorial pier.

please; urethane floor-and-deck enamel is probably the most durable and readily available coating.

CONSTRUCTING EQUATORIAL HEAD ASSEMBLIES

The equatorial head now consists of a pile of wood and hardware. Label both sides of each piece. Lay out the holes and countersinking for both sides. Below is a checklist of the parts:

A: 10″ diameter, three ⁵⁄₁₆″ holes and three ⅜″ holes for three pairs of push-pull bolts, 2½″ center hole

B: 10″ diameter, three ⅜″ holes for tee nuts matching the pull-bolt holes in A, ½″ center hole

C: 10″ diameter, Formica-faced, three 1″-diameter countersunk areas for clearance of B's tee nuts, ½″ center hole

D: 10″ diameter with boss, Formica-faced, four countersunk areas for clearance of carriage bolts in E, ½″ center hole

E: 10″ diameter with boss, four ¼″ holes for carriage bolts, ½″ center hole

F: 6″ diameter, Formica-faced, 2½″ center hole

G: 6″ diameter, Formica-faced, 2½″ center hole

H: 10″x12″ rectangle, ten ¼″ holes for attaching telescope, ring of ⁵⁄₁₆″ holes for tee nuts for attaching 1½″ pipe flange.

W1: 5⅞″ diameter, ⁵⁄₁₆″ center hole

W2: 5⅞″ diameter, ¼″ center hole, ring of ⁵⁄₁₆″ holes for tee nuts for attaching 1½″ pipe flange

W3: 6⅛″ diameter, 2½″ center hole

When these pieces are complete, construct four major assemblies from them, as follows:

Polar Plate Assembly: Clamp parts A and B together with the push-pull holes lined up. Install ⁵⁄₁₆″ tee nuts in piece B; then glue and screw B and C together, taking care to align their center holes. Attach small mending plates where push bolts will bear. Round the ends of two ⅜″ dowels 1¼″ long, then glue them into the holes in part A.

Polar Tee Assembly: Bevel one edge of pieces D

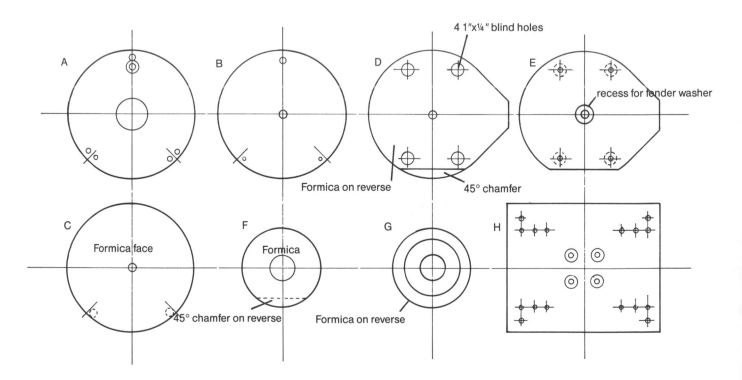

Parts for the equatorial head assemblies.

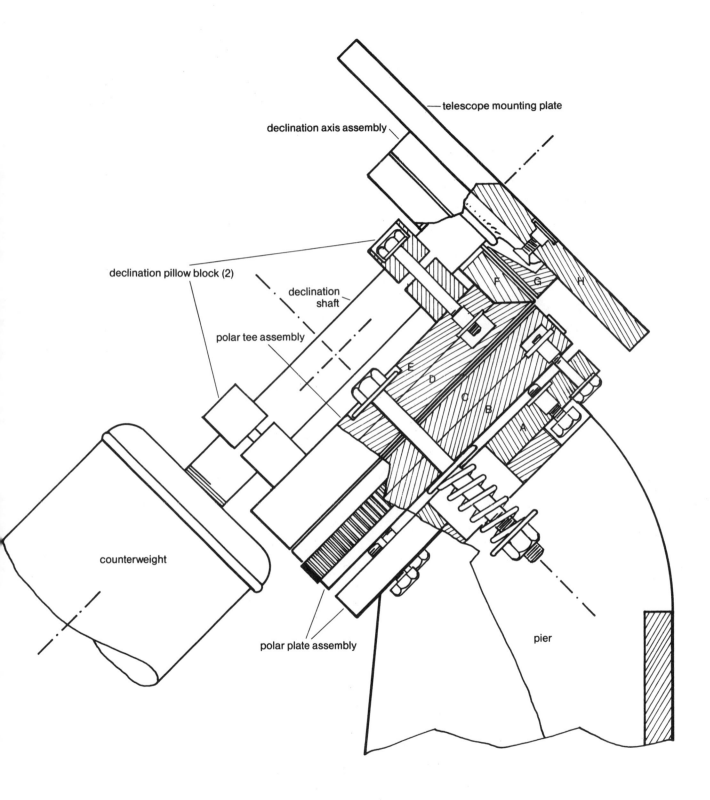

telescope mounting plate

declination axis assembly

declination pillow block (2)

declination shaft

polar tee assembly

counterweight

polar plate assembly

pier

Cross section of the equatorial head.

Bearing surfaces in the polar tee assembly must meet at right angles.

Declination bearing parts G and H; holes permit tightening the bolts holding the hidden pipe flange.

Declination assembly shown partially assembled, with the declination pipe in its pillow-block bearings.

and F to 45° so the outside edges meet at exactly 90° to each other; then glue these surfaces together. Clamp them so they remain perpendicular. When the glue has set, cut one side of part E flat so it fits tightly against both D and F and its center hole aligns with the hole in D. Place four carriage bolts into its holes, and glue and screw E to D and F.

Declination Axis Assembly: Bolt the floor-flange pipe fitting to the tee nuts installed in H; then chisel material from the backside of G so it fits over the flange. Glue G to H and clamp them together firmly. After the glue has set, thread the 12″-length pipe into the floor flange and run it up tight.

Counterweight Assembly: File or surform the smaller counterweight circles, W1 and W2, until they fit tightly inside the rim of the coffee can; then work the larger piece, W3, until it is just larger than the rim. Shape the outer edge of W1 into a curve that blends with the can's profile; then do the same to the larger piece, W3. Drill a 5⁄16″ hole through the bottom of the can. Install a ¼″ tee nut in W3 and bolt the floor

flange to W2. Chisel away excess material until W3 fits over the floor flange; then glue and screw W3 to W2. When the glue is dry, slip W1 into the bottom of the can, insert a ¼″-20 threaded rod into the tee nut in it, then fit the threaded rod through the hole in the center of W2 and W3, and, finally, slide them into the top of the can. Cut the threaded rod so it protrudes ½″ from the wood; thread a wing nut onto it to pull the wooden rings tight against the ends of the can.

CONSTRUCTING THE DECLINATION AXIS

Cut four 6″ lengths of 1½″-square clear pine, and drill 5⁄16″ holes through each of them (these pieces should slip down the carriage bolts protruding from E). Designate two as "bottom" pieces and two as "top" pieces. Slip the declination axis (only the pipe) through the hole in F and measure the distance between E and the pipe. Using first a keyhole saw, then a large rasp,

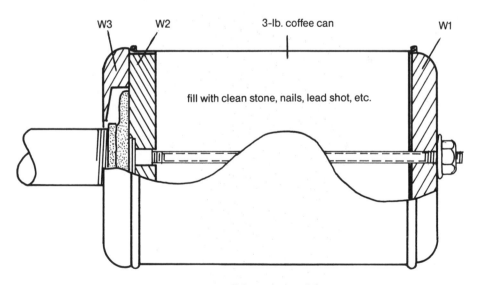

W3   W2   3-lb. coffee can   W1

fill with clean stone, nails, lead shot, etc.

Cross section of the counterweight.

Major subassemblies of the equatorial mounting before finishing.

remove a semicircle of wood from the two bottom pieces until, when they have slipped onto the carriage bolts, the declination axis pipe passes through the center of the hole in F and remains parallel to the surface of E. Repeat this for the top pieces so they too fit the pipe. These wooden parts form an adjustable "pillow-block" bearing made of wood. Bore a ¾″ countersink in them to expose the tops of the carriage bolts.

Cut a sheet of medium emery cloth to fit around the declination pipe; then put the pipe; then put the pipe through the pillow blocks. Turn the declination axis slowly, by hand, until the emery cloth has worn the contact surface of the pipe smooth. Remove the pipe and clean off any remaining grit. Replace it in the bearings, but this time, instead of emery cloth, line the bearings with felt. Add a drop of oil to each bearing; then turn the declination axis until the abraded sections of the pipe become shiny. Remove the felt pads. Cut small blocks of hardwood and glue them to the bearings. Tack bits of Teflon to the top bearing sections.

Prime all exposed *wood* parts of the assemblies. Do not prime or paint Formica surfaces or the locator dowels in part A. Fill, sand smooth, and paint the parts with two coats of urethane floor paint.

The top and bottom declination bearing blocks are made of white pine; optional maple inserts provide longer wear.

ASSEMBLING THE MOUNT

With the major parts completed, it remains now to assemble them into a complete working telescope mount. Begin by gluing and screwing part A firmly to the top of the pier. Run six #8x2″ screws into the pier side pieces and another six #10x1½″ screws into the top plate. Check to see that the polar plate assembly fits on the top of the pier.

Cut 6″ and 10″ circles of ¹⁄₃₂″ Teflon (if you can't find sheet Teflon, substitute thin Mylar or acetate plastic). Cut 2½″ and ½″ center holes in them, respectively. Place a ½″ hex-head bolt through the central hole in the declination tee assembly, through the Teflon, and through the polar plate assembly. Drop a fender washer over it, then the stiff spring, then another fender washer, and, last, a nonslip nut. Tighten the nut until the polar plate and the tee assembly pull together firmly.

Place the 6″ Teflon circle over the declination pipe; then slip the declination axis through its wooden pillow blocks and bolt the top pieces on. Attach the equatorial head to the pier, then run on the counter-weight. Move it around experimentally, tightening and loosening the bearings until it moves smoothly but not freely. There isn't much more to do until the telescope and cradle are sufficiently complete to attach the declination plate. At that time, fill the counterweight with crushed stone (or any other dense, inexpensive material), jam any remaining space with cloth or newspaper, then reattach the counterweight. Final adjustments must wait. Finding the optimum tensions means experimenting with an operating telescope.

For slow motion in right ascension, drill and attach metal plates to the boss on the declination tee assembly, insert a ¼″ rod through it with a knob on one end and a timing gear on the other, then glue the timing belt to the side of the polar plate. The slow motion can be added at any time, even after the mount is completed.

The optional right-ascension, slow motion assembly allows you to follow the stars by turning the large knob. It uses a plastic gear and rubber timing belt.

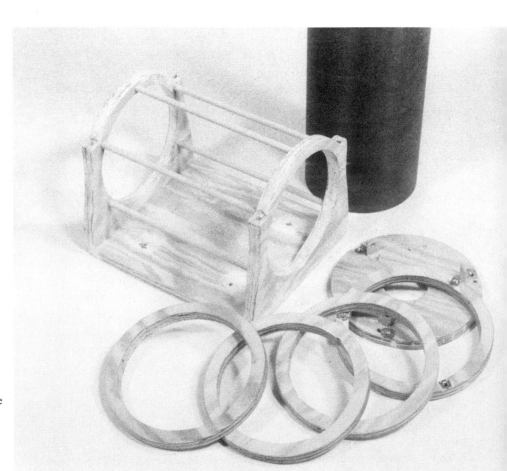

Tube rings, the mirror cell, the tube, and the tube cradle before assembly.

TUBE AND CRADLE PARTS

Before preparing any of the tube assembly, first obtain a suitable tube. One-eighth-inch phenolic (Bakelite) is nearly ideal—it is light, stiff, and easily worked—but spiral-wound paper tube works satisfactorily. Fiberglass works well but is expensive. The tube's length should at least equal the focal length of the mirror and preferably be 4″ to 6″ longer, with an internal diameter no less than ¾″ greater than the mirror's diameter, or 6¾″. The dimensions of the prototype are 7″ outside diameter by 50″ long.

Begin by laying out five circles on the remaining plywood. Each should have a diameter 1½″ greater than the outside diameter of the tube, or 8½″. Inside four of these circles, mark a 7″-diameter circle. These are to become the tube reinforcing rings; the remaining circle is the mirror cell. Cut them with a saber saw or a jigsaw, taking care that the cut remains perpendicular to the wood surface. Don't force the blade—this will make it bend. Cut the inside diameters of the four tube rings carefully. After cutting the first, check to see that it fits *snugly* over the outside of the tube. If the fit is loose, make the next cut smaller.

Next, cut the two rings and the base of the cradle assembly. The inside radius of the circles must be slightly larger than the inside diameter of the four tube rings because the tube must be free to turn inside the cradle. The cradle base is a simple rectangular piece.

Clamp the rings and cell together, then surform and sand them to a common outside dimension. Remove the cell, clamp the four tube rings together, and smooth the inside cuts. Do not open the rings too much. They must be snug on the tube even if they're a little ragged. Do the same for the cradle rings but work toward a ⅛″ greater inside diameter than the outside diameter of the tube. Because these parts are thin and relatively delicate, plug any voids with a mixture of wood glue and sawdust (which is stronger than cellulose filler), let them dry, sand them smooth, then apply a coat of primer/sealer before doing any more work on them.

CONSTRUCTING THE MIRROR CELL AND TUBE

Start work on the tube by clamping the best of the tube rings to the mirror cell and drilling holes for

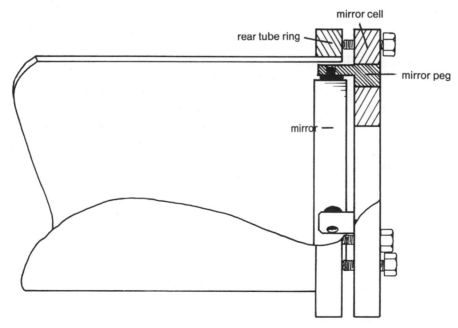

Cross section of the tube end with the mirror and its cell.

three equally spaced pairs of push-pull bolts in them. For long-term stability, cement the tee nuts in place with an epoxy cement.

Next, make the mirror cell. Drill three ¾″-diameter holes through the mirror cell on a diameter ¼″ greater than the diameter of the mirror. These holes will hold the mirror pegs. Make three of these pegs from 2″ lengths of ¾″ wooden dowel. Cut each dowel across a diameter, longitudinally 1¼″, then cut away one side. Drill a ³⁄₁₆″ hole through the half-section of the dowel ½″ from the end, then open it with a countersink bit from the flat side. Glue the holders into the holes in the mirror cell with their full diameters flush with the surface. When the glue dries, countersink three holes on a radius of 2¼″ in line with the mirror holders; then remove any excess dowel protruding from the back of the cell, fill, sand the cell smooth, and apply a coat of primer/sealer.

Next, determine where the focuser will be. If you made it yourself, you already know the focal length of the mirror; if you bought the mirror, you'll find the focal length inscribed on the back or side of the glass. The distance from the back of the tube to the center of the focuser hole equals the focal length of the mirror *minus* half the diameter of the tube *minus* the racked-in height of the focuser *minus* 1″ of focus-out allowance *plus* the thickness of the mirror.

Getting this right is critical: Double-check your math and make sure you understand the rationale for doing this (see Chapter 2) before making any cuts. Mark this and the holes for the bolts that will hold the focuser. The spider bolt holes must be separated from the focuser hole by the longitudinal distance between the center of the diagonal mirror and the bolts on the ends of the spider's legs. Measure the diagonal holder and spider carefully, then mark the locations for the holes. Cut the tube to length with a saber saw or a handsaw; use a hole saw to cut a 1½″-diameter hole for the focuser.

Attach the tube rings next. After drilling and countersinking eight radial pilot holes for a #8 wood screw in the middle two, slip them on and leave them loose. Slip the front and back rings on, drill eight evenly spaced pilot holes for #4x½″ wood screws from the inside, slip the rings off, smear their inside diameters with glue, slip them on again, and screw them in place. When the glue is dry, fill any voids between the rings and the tube with filler. If the tube is phenolic, spiral-wound paper, or concrete-form tube, coat it with primer/sealer (working around the two loose rings).

The next step is to paint all parts. A flat-black enamel with a coarse surface texture works best for the end rings and tube interior. You may opt to spray

Small wood screws and glue secure tube end rings.

To ensure that edges are sealed, cut all openings (such as the focuser hole) before you coat the tube with primer/sealer.

the inside of the tube; apply three or four light coats rather than two heavy coats. Paint the mirror cell similarly, but between each coat, burnish the cell with a stiff brush to remove any loose paint. Paint the exterior of the tube in sections, working around the loose rings. Two coats of marine enamel or urethane floor-and-deck paint form a lasting surface. You may paint the rings also, but they will require a touch-up coat after being attached to the tube. Allow the paint to dry thoroughly in a warm place, then brush the interior surfaces to remove all poorly adhered paint and dust.

Finally, install the mirror and diagonal in their respective cells. Buy a fresh tube of aquarium cement or bathtub-caulk silicone adhesive. Prepare a clear work area, with the cell at its center. Place the mirror faceup beside the cell. Set three 16d nails on the cell to serve as spacers. Lower the mirror onto the nails to check that nothing binds, then remove it. Squeeze three ¼″-thick, ½″-diameter blobs of adhesive into the countersunk holes. Lower the mirror into the cell—it will squash the blobs. Squeeze adhesive through the holes in the holders until it appears around the edges of each holder (don't let it touch

Before lowering the mirror into the cell, place three blobs of silicone adhesive in the indents in the cell. Nails serve as spacers.

Silicone adhesive injected through the mirror support posts complete installation of the mirror in its cell. (above)

The optical parts in their holders before being placed in the tube. (below)

the aluminized surface). Allow the adhesive to cure for 24 hours; then withdraw the spacing nails.

Assemble the diagonal holder according to the maker's directions, but squeeze a thin bead of glue around the back side of the mirror. This ensures that the secondary will not become loose or shift.

Now determine the tube's balance point. Bolt the focuser to the side of the tube, add the finder rings and finder at a convenient spot, and temporarily install the primary and secondary mirrors. Slide the loose tube rings to an equal distance on either side of the estimated balance point. Place an eyepiece in the focuser, then balance the tube across a large dowel. Check to see that the loose tube rings are each an equal distance from this point, adjust them, and find a new point if necessary. Mark this point, then remove all delicate parts from the tube again. Store the primary mirror and cell, diagonal mirror, holder, and spider assembly in a clean place until final assembly.

The tube is now nearly finished, but the remaining steps require that other parts be completed. The cradle must be fitted to the tube, and then the tube rings can be attached. After that, the tube exterior can receive its final coat of paint and be ready for you to install the optics and test the instrument.

CONSTRUCTING THE CRADLE

Construct the cradle as you work on the telescope tube so that when you finish it, you can go right ahead and complete the telescope. In addition to its base and rings, you'll need for the cradle 5′ of ⅜″ dowel rod, six #8x1½″ flat-head wood screws, four ¼″-20 internally threaded inserts, and four ¼″-20x3″ hex-head bolts. (If you can't find threaded inserts at your local hardware store, order them from a hardware catalog [see Appendix B]; if you can't find them anywhere, substitute mending plates.)

Drill holes for and insert four ¼″-20 tee nuts in

Drill bolt holes for reassembling the cradle before cutting it.

the cradle base plate. These must match holes in the declination plate, part H of the equatorial head, so check to see that they do at this point.

Decide which is the better side of each cradle ring; then clamp the bad sides together and surform or plane them to the same dimensions. At each of the five locations for the connecting dowel rods, drill a single ¹⁄₁₆″ hole through both pieces, then separate them and drill ½″-deep ⅜″ holes for the dowels. Temporarily nail the rings to the base piece and drill pilot holes for wood screws. Cut and fit five lengths of ⅜″ dowel to fit between the cradle rings. Finally, if you intend to use threaded inserts, drill a ¼″ hole 3″ deep through the flat side of each ring; if not, do not drill these holes.

Knock the cradle parts down, apply wood glue to all the mating surfaces at one end, then assemble that end. Do the same for the other end and assemble it too. Run wood screws into the base through both rings, measure the exact distance between them, then clamp the upper ends of the rings, keeping that same distance between them. This ensures that the tube will rotate smoothly in the cradle. After the glue has set, go over the whole assembly carefully, filling gaps and weak spots in the rings with a mixture of sawdust and wood glue, and allow this filler to set. At this stage, the cradle should feel light, strong, and rigid in your hands. Sand it smooth, fill small voids, then sand it again and apply a coat of primer/sealer.

The next step is to cut the cradle into two parts and then rejoin it. Clamp one ring to another piece of plywood of about the same size and saw it through from the inside; then clamp the other piece to another piece of wood and saw through that. Unclamp the assembly and remove the upper cradle. Drill each of the four holes through the sides of the rings to the body diameter of the threaded insert, then install them. Bolt the upper cradle back on with 3″ hex bolts.

Side-on and end-on views of the assembled cradle.

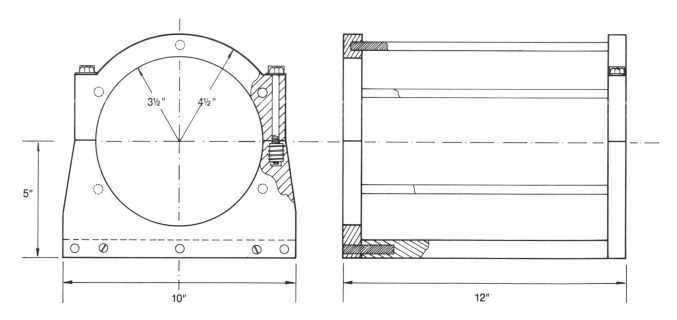

3½″   4½″

5″

10″

12″

(If you cannot find threaded inserts, screw a ½″x1½″ mending plate to the inside surface of each cut. This will accomplish the same end, but it is not as elegant.)

Disassemble the cradle and clamp the lower half to the bench. Place the telescope tube in it with the balance point in the center of the cradle. Bolt on the upper cradle. Slide the loose tube rings up to the cradle rings, drill pilot holes in the tube, then screw the rings in place.

Cut four strips of felt to fit between the cradle and tube for both lower and upper cradle halves, and two more pieces of felt to fit between the lower cradle and the tube rings. Attach these to the cradle parts with contact cement and press them firmly into place. When the cement has completely dried, paint the cradle and telescope.

The cradle is sawed to permit installation of the tube.

The cradle ready for use—even the felt strips have been glued to the completed unit.

ASSEMBLING THE TELESCOPE AND COLLIMATING ITS OPTICS

At last, your telescope is nearly ready to go! Place the lower half of the cradle on a firm worktable and lower the tube in it. Bolt on the upper half of the cradle, running the bolts in tight enough to restrict but not prevent the tube's rotation. If the paint is still slightly soft, wait an extra day or two—you'd be pretty unhappy if the felt stuck to the paint.

Attach the finder rings and any other hardware, handles, or gadgets you want to the outside of the tube. For the final time, brush out dust, debris, chips, or loose washers. Once the optical parts have been installed, you'll want them to stay clean.

Bolt the mirror cell to the end of the tube with the push-pull bolt pairs. The push bolts should hold the cell approximately ¼″ from the tube ring, and the pull bolts should be moderately tight. Do not force them. Look down in the front end of the tube: If the reflection of the open end of the tube and your face are not approximately centered in the mirror, adjust the push-pull bolt pairs until it is. This step places the optical axis of the mirror *roughly* parallel to the mechanical axis of the tube.

Assemble the spider in place. If you must, bend its legs to get it in, but try not to bend them sharply. Tighten the spider leg screws so the spider legs make a middle-toned "bong" when you pluck them. Next, reach in and install the diagonal and its holder. Now look down the focuser tube—the center of the diagonal mirror should lie exactly in line with the axis of the tube. If it is too far forward or back, adjust the holder longitudinally until the mirror appears centered. Rotate the holder so you see the primary mirror reflected in the diagonal.

The mirror cell bolts to the bottom end of the tube.

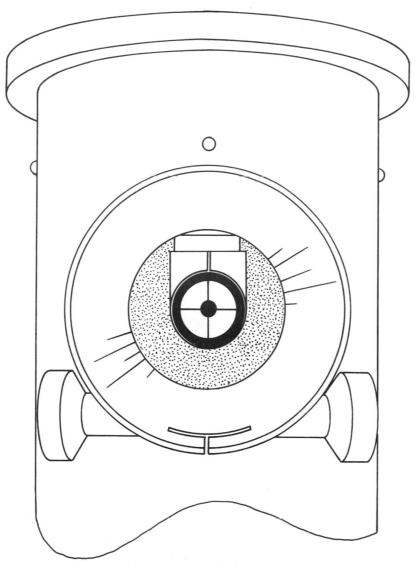

The appearance, looking down the focuser tube, of a properly collimated 6″ f/8 Newtonian reflector.

Center the reflection of the primary in the center of the diagonal by rotating it about its axis and tilting it. This is a difficult job since you will be working mostly by feel at some distance inside the tube, and the adjustments on some commercial holders tend to slip out as soon as you think you've got everything right. Persevere and eventually you'll get it both right and tight.

Finally, adjust the primary mirror by means of the pairs of push-pull bolts until the reflection of the diagonal is exactly centered in the primary. With this cell design, the alignment of the primary goes quickly;

and once it has been adjusted, the adjustment should stay put.

If the mount is finished, bolt the cradle to the declination plate, balance the instrument about the polar axis by adding or subtracting from the counterweight, then take the telescope outside for a trial run. The most devastating error at this point is to find that you calculated the distance between the focuser and the back of the tube incorrectly and cannot bring a star image to focus at all. If you were careful when you did this calculation, the image of a star will focus about 1″ out from the focuser's full-in position.

After all the work you've done, your first night with your new telescope should be an enjoyable one. Ignore the inevitable minor problems. Even though you'll probably find that the polar axis should be a little looser or a little tighter and that the declination may stick a bit, tonight's a night for looking. Spend the second night in getting everything tuned up right.

If you made the mirror, of course you will have tried it all before, with the unaluminized mirror taped temporarily into the mirror cell. Tonight, though, the image formed by the aluminized mirror is 25 times brighter. Your favorite first-magnitude test star will seem unbearably bright. Give in to the temptation to search for some faint galaxies.

A note on finishing: You may find yourself suffering "postpartum" blues when your telescope is done, especially if suddenly you face an unfamiliar sky full of unknown stars. Don't let it happen. Each night after you saw the last cut, drive the last screw, or brush on the last paintstroke, go outside and study the stars for five or ten minutes. Learn their names so that when you're ready to view them with your telescope, you'll feel that you're meeting old friends.

Close-up showing optional right ascension slow motion installed on the polar disk.

The completed telescope is both elegant and functional.

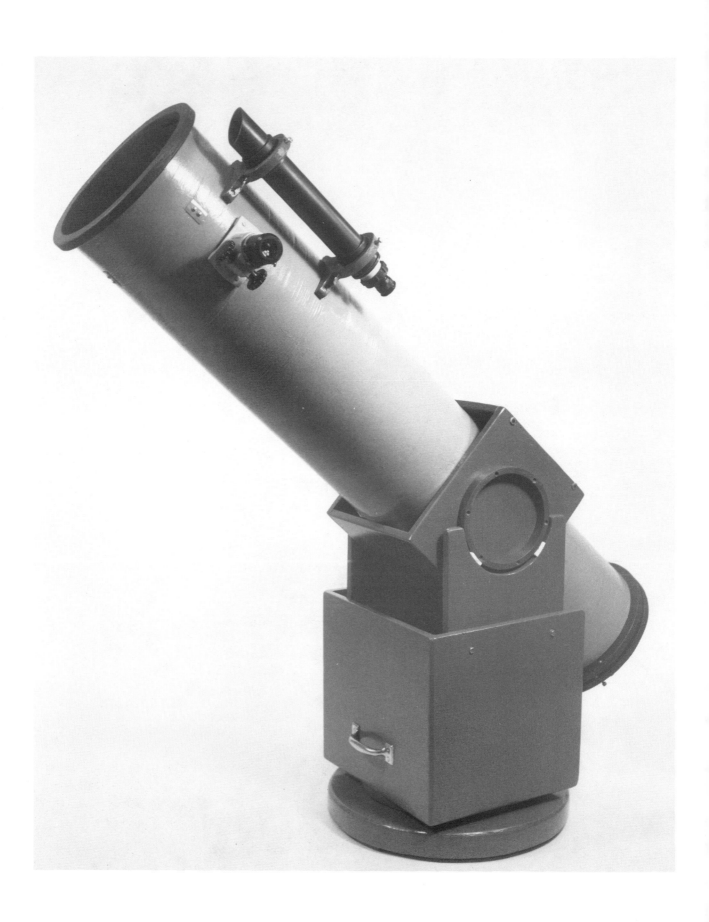

Building a 10″ Dobsonian Reflector

7

A 10″ reflector is capable of revealing thousands of star clusters, nebulae, and galaxies on clear, dark nights, and of resolving exquisite lunar and planetary detail on nights of good seeing. This telescope's Dobsonian mount allows portability and quick setup and is suitable for both casual observing and serious, long-term programs, such as observing variable stars.

If you want to make its primary mirror yourself, be warned that figuring a 10″ mirror to a high standard of performance is not an easy task, especially if it's to be your first mirror. If you've already built one working telescope, you know the pleasure that making a fine mirror brings and will find the patience to finish the job.

The mount is a classic Dobsonian, perfect for getting started in large-aperture observing. If you wish, install rollers on the base for "roll-'em-out" style observing; just be sure to raise the wheels off the ground for stability while actually observing. But remember that your telescope is not limited to *this* mounting only. You can place the tube on a permanent equatorial for astrophotography or high-power observing, then return it to its Dobsonian mount for observing trips to dark mountain or desert skies.

PLANNING THE JOB

The mechanical construction of the telescope will take two to three weeks of diligent evening work. The materials required are:
- one full sheet of ¾″ plywood
- a slightly rough Formica-faced particle-board sink cutout
- a 6′ length of 12″ Sonotube or other concrete-form tube
- two 8″-diameter rings of PVC pipe
- a 16″x17½″ piece of ¼″ plywood
- a few quarts of sealer and paint
- an assortment of screws and other hardware
- a focuser (for either 1¼″ or 2″ eyepieces, your choice)
- a spider for a 12″ tube
- a 2.5″ minor axis diagonal holder
- a finder telescope (preferably at least 8x50mm)
- as many eyepieces as you want

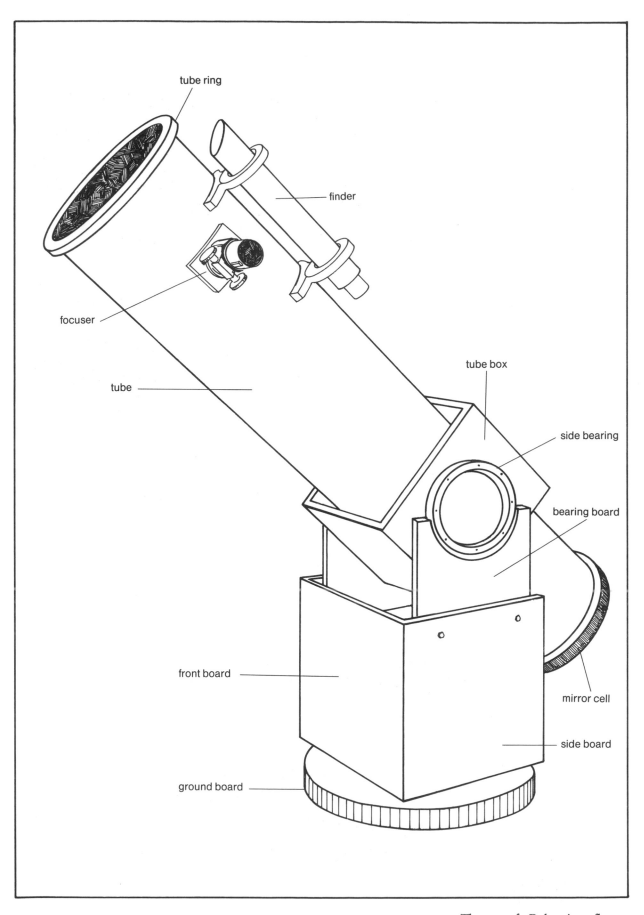

tube ring

finder

focuser

tube box

side bearing

bearing board

tube

mirror cell

front board

side board

ground board

The parts of a Dobsonian reflector.

As with all telescopes, you can put as much or as little effort into building the accessory parts as you care to.

Begin by ordering the optics; then start accumulating the more exotic materials. Concrete-form tube is not the best possible tube material, but it is by far the least expensive—about $10 for a 6′ length. Substitute phenolic plastic or fiberglass if you prefer, but tubes of these materials may cost over $100. Spiral-wound paper is less expensive, but concrete-form tube *is* spiral-wound paper too. Appendix A, on materials, gives some tips on removing the waxy outer coating that some tubes have. Start stripping the wax right away; a tube that cannot be dewaxed should be replaced.

For the side bearings, obtain two 1″-long rings cut from 8″-diameter, ½″ wall-thickness PVC pipe or any other smooth, cylinder items 7″ to 10″ in diameter and 1″ to 1.5″ thick. Metal film cans work well, as do discarded 45-rpm record-player turntables, molded plastic plumbing parts (although these are usually smaller than 7″, they'll work), and metal skillet lids. Use some imagination—the possibilities are endless. If you can't find something ready-made, cut plywood circles and attach an aluminum, stainless-steel, or Formica edging to the periphery with contact cement.

Review Chapter 9 and decide if you want to make or buy the focuser, diagonal holder, spider, and finder. Making them is more satisfying if you enjoy crafting neat, precise machine parts, but if your interest lies solely in getting a telescope "on the air," buying them is the expedient route. If you plan to buy them ready-made, it's a good idea to order them well ahead.

Of course, also obtain Teflon for the bearings. Buy enough for four side bearings ¾″x1½″x¼″ thick, two lateral side bearing pieces ¾″x¾″x¼″ thick, and three bottom bearing pieces 2″x2″x¼″ thick. If you cannot locate ¼″-thick Teflon, ³⁄₃₂″ or ⅛″ will work, but thinner bearing pads are harder to attach securely.

The optical parts for this telescope are a 10″ f/6 paraboloidal primary mirror, a 2.5″ minor-axis elliptical diagonal mirror, and eyepieces. F/6 is nearly ideal as an all-around focal ratio: The telescope comes out at a convenient length; the image scale is reasonably large; and the focal ratio is large enough that most eyepieces will give good image quality. F/4.5, f/5, f/5.6, and f/8 mirrors are available, but remember that

substituting these alters the length of the tube and the height of the rocker-box side boards. If deep-sky observing is your goal, pick a shorter focal-ratio mirror; for planetary and lunar observing, pick a longer focal ratio. Order early: Although 4″ and 6″ mirrors are usually stock items, 10″ mirrors may not be. Depending on the supplier, delivery may take from two weeks to six months.

If you're doing the optics yourself, plan on having both the telescope and the mount near completion in time for star-testing the optics. Testing can be done before anything has been painted; this allows for changing the focuser position to match the mirror's focal length. While the mirror is out to be aluminized, finish up the cosmetics.

LAYING OUT THE WOODEN PARTS

Purchase a full sheet of ¾″ A-C exterior-grade plywood for the 10″ telescope and mounting. Place the plywood good-side-up on a work surface and mark it for cutting. Several dimensions depend on the tube material. For the inside diameter of the reinforcing rings, use the outside diameter of the tube minus ¹⁄₃₂″. Since the inside diameters will be sanded before assembling the tube, cutting them a bit too small gives some sanding allowance. The base rings are not critical except that the outside diameters of all three must have exactly the same radius.

The dimensions of the tube box depend on the outside diameter of the tube. Measure its circumference to the nearest ¹⁄₁₆″, divide by pi (3.14) to get the diameter, then lay out the top and bottom pieces of the tube box ¼″ *shorter* than the tube diameter. The side pieces should equal the outside diameter of the tube plus twice the thickness of the plywood. The dimensions shown in the plans apply to a concrete-form tube with a 12″ inside diameter and an outside diameter of 12½″, yielding top and bottom piece lengths of 12¼″ and side piece lengths of 14″. Making the tube box slightly too small ensures that the box, when assembled, will hold the tube firmly.

The rocker-box side pieces and the base are 16″ square. The circle cut from the side bearing supports

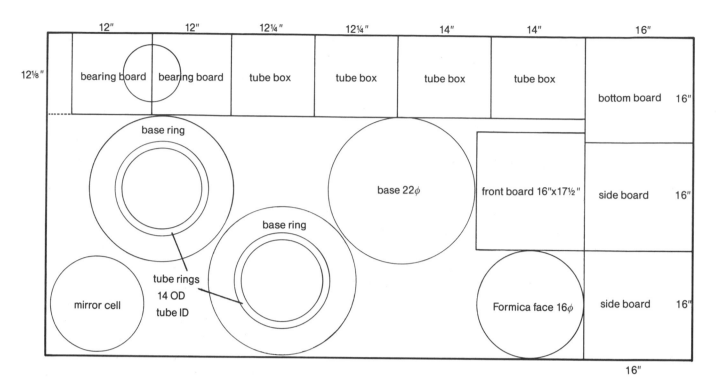

All the parts may be cut from a single sheet of ¾″ plywood, although the front board can be cut from ¼″ plywood and the bottom bearing from a Formica-faced sink cutout.

should have a radius ³⁄₃₂″ greater than the final radius of the side bearings. If the bearings are plywood circles, include the thickness of the covering material when figuring their radius.

When you've finished marking the ¾″ plywood, lay out the front board on ¼″ plywood. It should be 16″ high by 17½″ wide. This piece is made from ¼″ plywood only to save weight—any thickness will do.

Finally, scribe a 16″-diameter circle on the Formica-faced particle-board sink cutout. Mark the center so you can locate it again later. If you cannot find a sink cutout, cut an additional 16″-diameter circle from ¾″ plywood and bond Formica to it with contact cement. Formica with a very slightly rough surface is better than the shiny, smooth kind. Avoid textured types.

SAWING PARTS FOR THE TUBE AND MOUNT

At last, it's time to raise some sawdust! Support the plywood sheet on a solid work surface, then cut off the 16″ of the end that has the side boards and bottom board, and, next, the 12″-wide piece with the tube box parts. These two cuts reduce the difficulty of handling the remainder. A handsaw is entirely adequate for making these cuts.

Next, saw the side and bottom boards apart, then saw out the tube box parts. Cut the two side bearing supports apart with a scroll saw or an electric saber saw, taking care to obtain a smooth curve on each support. Remember that it's the support, not the circle, that's going to be part of the telescope.

The big base circle, rings, and tube-reinforcing rings now remain. Cutting large circles with a coping saw is tedious; an electric saber saw will do the job in the most efficient way. If you have a well-equipped shop, a router produces much smoother, more nearly perfect circles. Start the cuts at the inside and work out.

As soon as the first tube end ring is cut, check to see that it fits tightly over the end of the tube. If it's loose, it's too large and won't provide proper support for the tube.

Finally, cut the ¼″ plywood to dimension and, using

the saber saw, cut the 16″ bottom bearing circle out of the sink cutout. At the end of a single dusty session, you'll have made one base circle, two base rings, one bottom bearing, one rocker bottom, two rocker sides, two side bearing supports, four tube box parts, the mirror cell disk, and two tube reinforcing rings. Whew!

Pair up complementary parts, clamp them in a large vise or hang them over the worktable, and rasp, surform, and file them into matching pairs. Round off the outside edges, and fill any openings or voids in the plywood laminations with a thick mixture of wood glue and sawdust. After the glue mixture has dried thoroughly, sand these repairs smooth.

The Dobsonian telescope's rocker and tube box—before assembly.

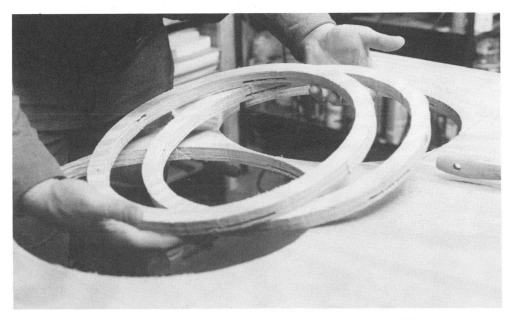

Do not expect plywood parts, such as these tube rings, to look finished immediately after being cut.

Surform or plane the tube box parts flat, square, and to the same width.

Surform or plane the mirror cell and tube rings to the same outside diameter.

CONSTRUCTING THE TUBE

The telescope now looks more like a pile of construction leftovers than anything else. At closer inspection, however, the collection of telescope hardware is rather imposing. Whatever is being built, it's pretty unusual!

If you have not yet removed all the wax from the tube, do so now. The most effective technique depends on the type of tube, and so it varies greatly. Don't be afraid to experiment on the extra foot or so of tube. If you use solvent to remove the wax, work outdoors with plenty of ventilation, and wear gloves. When the tube is clean, let it dry outdoors, preferably in the sun, for several days.

Begin work on the tube by deciding on its length. A safe and reasonable tube length for visual observing is the same as the focal length of the mirror; this is the case with the prototype, which has a 60″ tube for a 60″ focus mirror. However, for optimum baffling of stray light—especially important for long-exposure, deep-sky astrophotography—the tube should be at least one tube diameter longer than the focal length of the mirror.

Cutting the tube perpendicular is important if for no other reason than looks. Wrap a 12″-wide strip of paper tightly around the tube—ordinary household waxed paper is fine. Its edge should define a diam-

Plan of a push-pull bolt pair for ¾″ plywood and ¼-20 bolts and tee nuts.

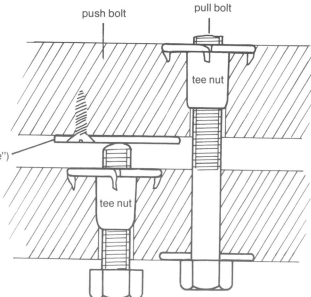

push bolt

pull bolt

tee nut

pressure bearing plate ("mending plate")

tee nut

eter. Mark the tube, then cut it. A sharp hardware knife cuts concrete-form tube cleanly, but a fine-toothed saw will do well too. Check to see that the cut end is perpendicular with a carpenter's square, measure the tube length, wrap paper around, mark, then cut the other end.

Deburr the cut ends with a sharp knife, smooth them with sandpaper, then go over the whole length of the tube, sanding down any scuffs, bumps, or high spots. Once again move outdoors and coat the tube with a generous coat of primer/sealer inside and out. (Bolt the paintbrush to a 2′ length of wood to do the inside.) Soak the newly cut ends thoroughly. The runny sealer soaks into the paper, carrying a waterproof resin that bonds the paper fibers. The sealer will, in all probability, simultaneously dissolve any residual wax.

While waiting for the tube to dry, file and sand the inside diameters of the tube-reinforcing rings until they just barely slip over the tube scraps. Examine the rings for defects, then pick the better of them for the mirror cell ring. Clamp it to the mirror cell circle and drill pilot holes for three push-pull bolt pairs around the periphery. For each hole pair, drill two $\frac{1}{16}$″ holes through both pieces 1″ apart, centered in the ring. Separate the ring and cell. Mark one hole in each pair "pull" and the other "push."

From the back side of the cell, counterbore each pull hole $\frac{3}{4}$″ in diameter and $\frac{1}{8}$″ deep with a spade bit, then drill a $\frac{5}{16}$″ hole through. To the front side, screw half of a two-hole mending plate to reinforce the push location.

From the front side of the ring, counterbore each pull hole $\frac{3}{4}$″ in diameter and $\frac{1}{8}$″ deep with a spade bit, and do the same from the back side of each pull hole. Drill the pull holes through with a $\frac{1}{4}$″ drill and the push holes through with a $\frac{5}{16}$″ drill.

Install a $\frac{1}{4}$″-20×$\frac{3}{4}$″ tee nut on the back side of the counterbored holes of the ring and cell. Glue a $\frac{1}{4}$″ washer into the front side counterbore in the ring. From the front side, run a $\frac{1}{4}$″-20×1″ hex-head bolt into the push hole and a $\frac{1}{4}$″-20×1$\frac{1}{2}$″ hex-head bolt into the pull hole.

Since you're already at work on it, complete the mirror cell. Cut three 2$\frac{1}{2}$″ lengths from a $\frac{3}{4}$″-diameter dowel rod. Saw a bit less than halfway through it $\frac{3}{4}$″ from one end, then saw away half of the length of each dowel. Round the sharp edges. In the resulting flat surface, drill and countersink a $\frac{3}{32}$″ hole $\frac{3}{4}$″ from the end. At a radius $\frac{1}{8}$″ greater than the radius of the primary mirror (measure this—a 10″ mirror may really be 9$\frac{7}{8}$″ or 10$\frac{1}{4}$″ in diameter!) and in line with the push-pull pairs, bore three $\frac{3}{4}$″-diameter holes. Glue the round ends of the dowels into these holes. These dowels are mirror support posts.

The mirror cell is mounted on the tube ring by three pairs of push-pull bolts.

Three mirror support posts will hold the primary mirror; each post is in line with a push-pull bolt pair.

Primer/sealer is a clear coating that hardens wood and paper while sealing out moisture.

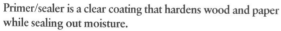

Both glue and wood screws hold the end rings to the tube wall.

One inch inward from each support post and in line with the post, drill a ¼″-diameter hole and countersink it. This anchors the adhesive that holds the mirror into the cell.

Meanwhile, the primer/sealer on the tube will have dried, so coat the insides of the rings with wood glue and slip them over the tube, flush with the ends. Double-check to see that you installed the back side of the cell ring facing the back of the tube. From the inside of the tube, drill and countersink pilot holes for six #6x½″ flat-head wood screws per ring, then run the screws in. When the glue dries, the rings will be securely attached. Fill dents in the tube or gaps between the rings and tube; then sand them smooth and recoat these areas with primer/sealer.

Now, where do the holes for the focuser and spider bolts go? First, calculate the distance from the desired focal plane to the center of the tube: the sum of the focuser height, a focus allowance of 1″, and half the outside diameter of the tube. This will work out to approximately 10″, but of course, it varies, depending on the components.

Before applying the second coat of paint, drill all openings in the tube.

Polyurethane floor enamel is a tough, water-resistant exterior finish recommended for all tubes and wooden parts.

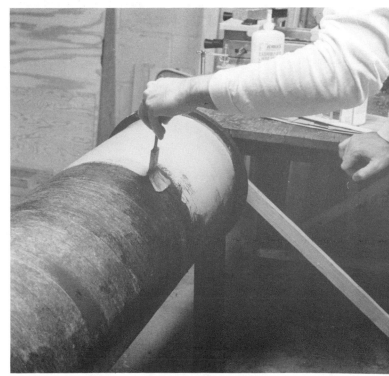

If the mirror's focal length is 60″ and the focal plane to the diagonal distance is 10″, the distance from the diagonal to the mirror is 50″. Since in this design the front surface of the mirror will lie 1½″ inside the back end of the tube, the diagonal should lie 51½″ from the back end of the tube, and the center of the focuser hole lies at this position as well. (See also Chapter 2.)

The distance from the center of the diagonal to the bolt holes for the spider also depends on the spider and the diagonal holder. Typically, this distance is 4″. Measure the spider, then add this distance to the focuser-hole distance—in our example, 55½″—some 4½″ inside the front end of the tube.

Use a hole saw to drill the focuser hole and a regular drill for the focuser mounting holes and spider bolts. It's convenient to drill the mounting holes for a finder telescope, the handles, and any other hardware at this time. Soak the holes with primer/sealer and let it dry thoroughly.

Finish the interior of the tube with two coats of flat-black paint—the flatter and blacker it is, the better. This is a messy job, so bolt the brush to a piece of wood (as for the primer/sealer) to make it easy on yourself. Coat the exterior of the tube with two coats of a good, tough paint such as polyurethane floor enamel.

ASSEMBLING THE TUBE BOX

The tube box is a simple unit made from four pieces of wood and two side bearings. Three sides of the box are permanently attached; the top is removable for inserting or, if necessary, removing the telescope tube.

Tack the two box side pieces (the longer pair) to one of the shorter pieces. Check to see that it's square,

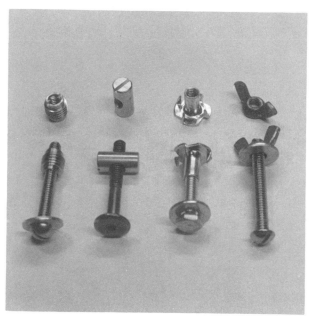

Special fasteners, left to right: threaded insert, right-angle joint connector, tee nut, and wing nut.

Clamp the side bearing to the tube box and drill pilot holes for wood screws.

then glue and attach the sides with eight #8x2″ flat-head wood screws, four to a side.

To attach the top piece, obtain four right-angle joint connectors from a specialty hardware store or by mail order (see Appendix B). If you can't get right-angle joint connectors, use #8x2″ flat-head wood screws instead. Tack the top piece in place, then bore appropriate holes perpendicular to and through the side pieces. Remove the nails before installing the fasteners.

Attach the side bearings, carefully centered, to the sides of the box. Use wood screws for PVC pipe, stove bolts and tee nuts for metal rings, glue and wood screws for plywood disks. Take care that nothing protrudes into the interior of the tube box that could mar the tube.

Cut a 12″x36″ piece of 4-mil polyethylene plastic. Wrap it around the tube and lower the tube into the box. It should fit, but tightly. Place the top of the box on the tube, install a few of the fasteners, lean heavily on the top to force it down, and slowly draw the fasteners down. This crushes the tube slightly—about ⅛″ on each side—but grips it well.

If the box is too large and fails to grip the tube tightly, glue thin wood or cardboard spacers inside the box. If the box is too small, shave a layer of wood from the inside of the box so that it closes.

After filling, sanding, priming, painting, and allowing the parts of the box to dry thoroughly, put them together to recheck the tube fit. Don't forget the plastic: It prevents the paint on the surfaces from sticking and prevents digs and gouges in the tube. Having established fit, you're ready to put the telescope together.

INSTALLING THE MIRRORS

Brush and vacuum any dust and dirt from the tube thoroughly before installing the mirrors; keep them clean right from the beginning. Bolt the focuser and finder mounts in place; then touch up the exposed ends of bolts, washers, and nuts with flat-black paint.

Install the spider and tighten the leg bolts until the legs make a mid-toned "tung" when plucked. Install

Before assembling the tube box with wood screws, check to see that it fits the tube.

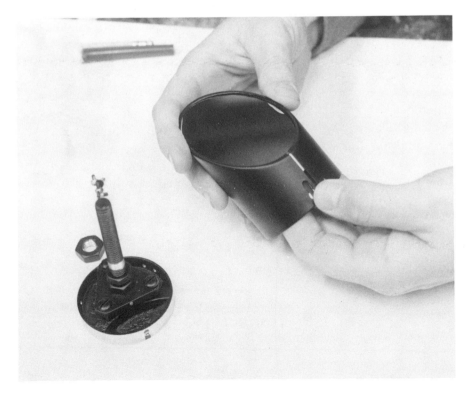

Insert the diagonal into its holder without touching the surface.

The spider and diagonal go into the front end of the tube; note the bolt holes.

Four bolts hold the focuser to the tube.

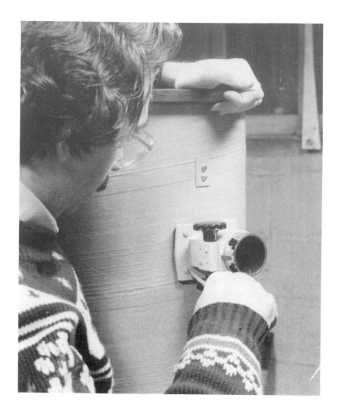

the diagonal mirror into the diagonal holder according to the manufacturer's instructions or, if you built the holder yourself, use your own plan. Install the diagonal holder in the spider in the tube.

If you're figuring the primary and need to test it on stars, set it between the mirror support posts, wedge a few sheets of cardboard between the posts and the mirror, then wrap a turn of masking tape over the posts and the side of the mirror. Bolt the cell to the cell support ring. Bring it into approximate collimation by centering the reflection of the focuser in the mirror. Place the telescope on the mount and locate a bright star. After collimation, examine and evaluate the extrafocal patterns; then bring the telescope inside, remove the cell, strip the tape off, demount the mirror, and continue figuring.

For the final mounting of the aluminized mirror, buy a fresh tube of silicone rubber adhesive. Place the mirror, the mirror cell, and three ³⁄₃₂″-thick spacers (16-penny nails are excellent) in a clean spot where they can sit unmolested for two days after you've glued them. Once you've picked the place, you're ready to do the gluing.

Place the spacers equidistant between the support posts. Into the three holes in the cell, squeeze three blobs of silicone adhesive roughly 1″ in diameter and ⅛″ thick. Gently lower the mirror between the posts and down onto the adhesive blobs, which should spread as the mirror contacts them; if it doesn't lift off the mirror, squeeze more adhesive onto the blobs, then try again. To improve the contact between the adhesive and the mirror, lift the mirror while rocking it slightly back and forth, then settle it on the spacers.

Shift the mirror slightly so that it's equidistant between the mirror support posts. Squeeze adhesive through the hole in each post until the adhesive expands to form a pad between the mirror and the post. *Do not allow any adhesive to get on the surface of the mirror.* After making a small pad between the mirror and each of the three posts, go around again, adding more adhesive until the pads are all between 1″ and 1¼″ in diameter.

Let the adhesive cure for 24 hours before doing *anything* to it, then remove the nails. Let the adhesive cure for another 24 hours before picking up the cell and mirror. These thick pads need several days' curing time to reach full strength and resilience.

While waiting, place a ⅛″-diameter drop of India ink

Silicone adhesive bonds the primary mirror into the mirror cell; allow 48 hours' curing time before disturbing the mirror.

Lowering the heavy mirror cell into place on the end of the inverted tube is a tricky and delicate operation!

Three pairs of push-pull bolts hold and position the mirror cell at the end of the tube.

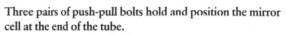

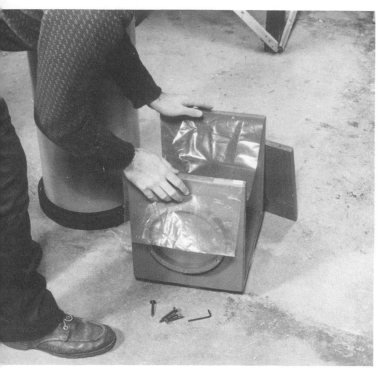

at the exact center of the mirror. This dot will help you see the mirror's center during collimation. Aside from this, leave the mirror alone.

When the waiting time is up, bolt the cell and mirror to the tube, leaving approximately ⅜″ between the cell and the cell support ring. Adjust the push-pull bolts so you can see your face reflected in the mirror when you look down the front of the tube.

The telescope tube is almost finished now. Before collimating the optics, it's a good idea to install the tube in the tube box. Add any and all remaining parts to the tube. Don't forget to put an eyepiece in the focuser. Locate the balance point, lower the tube into the tube box, slide the tube back and forth in the box until it balances perfectly, then put the top of the box on and tighten the fasteners.

From roughly a foot outside the focuser tube, center your eye on the axis of tube and examine the position of the diagonal mirror. Is it centered and in line with the focuser? If it's not, center the spider in the tube and adjust the holder longitudinally until it is.

Installing the Tube Step 1: Drape plastic over the tube box.

Step 2: Lower the tube into the box.

Step 3: Attach the fourth side of the box.

Step 4: Check and correct the tube balance.

A right-angle joint connector is the best way to attach the fourth side of the tube box.

Cut a disk from cardboard or plastic so that it just slips tightly into the focuser tube and punch a hole ¹⁄₁₆″ in diameter in its center. Slip this "centering disk" into the tube, then look through it. Adjust the tilt and rotation of the diagonal mirror so that the primary mirror appears precisely centered in the diagonal.

With the centering disk still in the focuser, adjust the primary mirror until the image of the diagonal mirror and the reflection of the little hole that you're looking through appear concentric in the primary. (Here's where that small black dot marking the center of the mirror really helps!) With rough collimation done, remove the centering disk.

If the mount is ready, place the telescope on it. Although the telescope may need a few adjustments, it should work well right from the start. Find a bright star, then align the finder on it. Put the finishing touches on the optical collimation by checking the extrafocal disks of a medium-bright star (see Chapter 11)—then get ready for the universe!

BUILDING THE MOUNT

The mount is straightforward carpentry. Begin by pairing the two opposite rocker sides against the rocker bottom board. Tack them on with a few finishing nails, then check for squareness. Make sure that the front board fits across the wide dimension of the box. If it is too wide, trim it to fit.

Tap the pieces about a ¼″ apart, inject wood glue, and drive the nails back in tight. Drill and countersink pilot holes for four #10x2″ flat-head wood screws on each side, then run in the screws. Turn the assembly back-side down; check to see that the sides are accurately perpendicular to the bottom. If they are not, force them inward or outward so they are.

Run a bead of glue along the three exposed front edges, then set the front board in place on them. Tack the front board to the front of the bottom board, then square up each side and tack it to the side. Drill and countersink pilot holes for 12 #8x¾″ flat-head wood

This view down the "business end" of the completed tube assembly shows the location of both spider and focuser.

The front board steadies the sides of the rocker box and keeps them perpendicular to the bottom board.

The bottom bearing is a disk of kitchen counter-top Formica, which slides on Teflon pads with buttery smoothness.

screws (four for each joint), then run in the screws.

Place the assembly upside down on a work surface. Center the 16″-diameter bottom bearing circle on the bottom board to check its fit all around. Remove it and place it Formica-side down, then turn the rocker over. Apply wood glue to the particle-board side of the bottom bearing, lower the rocker onto it, carefully center it, and tack it in place with a few finishing nails. Drill and countersink pilot holes for six #10x1¼″ flat-head wood screws, taking care not to drill through the Formica; then run in the screws.

When all the joints are thoroughly dry, round the corners, fill any remaining gaps in the box with wood filler, then sand it smooth. Round the corners, fill, and sand the bearing supports too. Determine the distance from the balance point to the bottom end of the telescope tube, then add 1″ for safe clearance. Clamp the bearing supports to the side boards, squared and centered, with their flat tops the same distance above the inside bottom of the rocker. Drill two ¹⁄₁₆″ pilot

holes through both side board and support on each side of the rocker, countersink holes for the washers ¾″ in diameter by ⅛″ deep, then enlarge the pilot holes in the support boards to ¼″ and those in the side boards to ⁵⁄₁₆″. Install ¼″-20 tee nuts in the side boards; attach the bearing supports to the side boards with ¼-20x1½″ hex-head bolts.

The base is 22″ in diameter and three layers of ¾″ plywood thick. The bottom two layers are rings, open in the center to reduce weight. Choose the better side of the 22″-diameter base circle as the upper surface, then glue and nail the two base rings to the bottom side of the circle. Glue three ¼″ plywood circles 2″ in diameter to the bottom rim of the base. These are the feet of the base. When the glue has dried, round the outside edge with a large wood rasp, then surform or plane it smooth. Fill any holes with wood filler and, finally, sand it smooth.

The last operation before painting is constructing the bottom pivot. This is a 3″x⅜″ hex-head bolt that

The base is assembled with glue and nails; the underside of the ring is coated with glue.

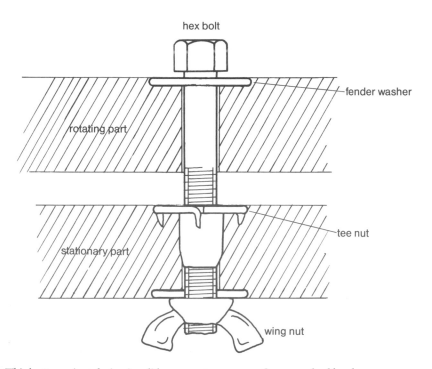

This bottom pivot design is solid, yet easy to construct from standard hardware.

Be sure to recess the heads of the brads half the thickness of the Teflon when you attach the pads to the base.

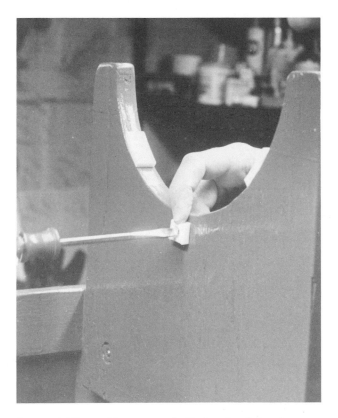

An extra Teflon pad between each side board and the tube box prevents unwanted side-to-side motion.

passes through a ⅜″ hole in the rocker bottom board and is bolted firmly into the base. Determine the center of the bottom board and drill a ⅜″ hole through it. Counterbore a 1″-diameter by ⅛″-deep recess in the top side of the base, and drill a ⁷⁄₁₆″ hole through the base for a ⅜″ tee nut. Pound in the tee nut.

Cut all nine pieces of ¼″-thick Teflon to size, bevel the edges, then drill and countersink holes for #4x½″ flat-head wood screws. Drill pilot holes in the bearing supports and base. (If you use thin Teflon, nail the pieces in place with small brads instead of wood screws. Be sure to sink the heads of the brads below the surface.)

This completes the mechanical construction of the mount. If you can't wait to try it out, or simply need to run some star tests on the mirror, screw the Teflon pads to the base and the bearing supports, bolt the supports to the side boards, and install the bottom pivot bolt. Plop the telescope into the rocker, and it's in business.

At some point, vacuum up all sawdust in the work area. Then finish the job by applying a coat of primer/sealer on all the surfaces of the base, rocker, and bearing supports. When it's dry, sand smooth with fine sandpaper and apply two coats of a good-grade paint to all surfaces. I recommend polyurethane floor-and-deck paint. It's extremely tough and durable.

Side boards are attached with hex-head bolts and tee nuts; this allows them to be repositioned if necessary.

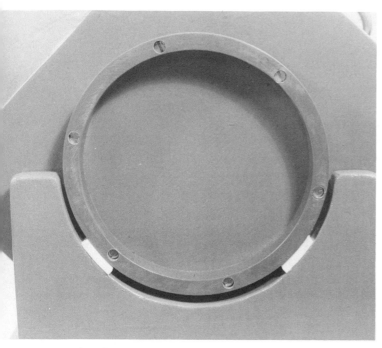

The side bearing rests on two Teflon pads attached to the side board; there is no play or wobble in it.

The hollow bottom side of the base need not be painted, but it must be sealed to prevent moisture damage.

TIPS ON USING YOUR NEW TELESCOPE

Your 10″ reflector is a powerful telescope, but if you're a novice at observing, you probably won't know how to exploit its capabilities. As emphasized previously, learn the sky *before* finishing the project. Learn the constellations first; then get practice with binoculars in exploring for clusters and nebulae.

If you've been observing with smaller telescopes, the light-gathering power of the 10″ aperture is impressive. It offers more than three magnitudes of light gain over a 2.4″ refractor, two magnitudes over a 4″, and one magnitude over a 6″. Globular clusters—dim, fuzzy, unresolved blurs with a 4″—will break into glorious, starry globes hanging in the blackness of space. Galaxies, hardly more than smudges with a 6″, present some detail to an observer with a 10″ telescope used faithfully under dark skies.

Of course the moon and planets will show features you can't see with smaller instruments. Of the telescopes described here, only the 6″ refractor, with a closed tube and superb optics, even begins to rival the performance of the 10″. If the mirror is a good one, this telescope will outperform the far more expensive 8″ Schmidt-Cassegrainians on the market and give an image that's half a magnitude brighter to boot!

However, a 10″ reflector is large enough that you must exercise care in its use. Allow, for example, a period between one and three hours for the telescope mirror to cool to ambient air temperature and form top-notch images. Storing the telescope on a cool porch or in an unheated garage keeps it closer to the air temperature and also makes it handier to set up.

The observing area becomes more important with increasing aperture. Avoid setting up on asphalt, concrete, or near to rooftops. Of course it won't harm the telescope if you observe while it's still warm or at a poor site, but you won't get all the performance it can deliver.

Optical alignment is far more critical for the 10″ than it is for the 4″ and 6″ telescopes, so don't neglect

At last—it's done! The telescope waits for its first night under the stars. (opposite)

On nights of good seeing, a 10″ reflector can deliver stunningly sharp views of the moon. Photo by Jean Dragesco.

it. You'll find directions for critical alignment in Chapter 11, as part of mirror testing.

Handling becomes more critical too. It is possible to drop a well-made 4″ telescope several feet without inflicting damage. But the mirror in a 10″ telescope is heavy enough that it may bounce out of its cell and be damaged if the telescope falls over. Six-inch reflec-

Deep-sky objects such as globular clusters come into their own only with relatively large telescopes. Photo by Walter Hamler.

tors tolerate long rides over rough roads; your 10″ should too, but it's best not to subject it to severe bumps and jars.

After a year or two of serious observing, you may wish to try your hand at astrophotography or another observing program in which tracking is essential. Quality clock-driven equatorials *are* within the abilities of an amateur telescope-builder. Reread the chapters on mounts and accessories; then broaden your horizons by exploring the wealth of ideas covered by the books and magazines listed in Appendix C.

The first telescopes were refractors. Galileo, Huygens, Cassini, Halley—even Isaac Newton, the inventor of the reflecting telescope—built and used refractors for fundamental astronomical discoveries. Refractors were the only telescopes for the first 150 years of telescopic astronomy.

Imagine a thin disk of glass—one surface convex and the other flat—a lens. As parallel rays of light from a star strike the front surface of the glass, the rays slow from the speed of light in the dense medium of the glass. Rays reach the center of the lens sooner than they reach the edges, so because photons must remain in step, they bend toward the axis of the lens. The incoming wavefront is no longer plane, but is convex toward the source of light as it passes through the glass. On exiting the back of the lens, and going from glass to air again, the rays are again bent toward the axis. The ray paths now move toward a common focus at which the light from the star ideally comes together as a tiny point.

Both refractors and reflectors focus light; they simply use a different means. Yet while all wavelengths are reflected equally, the amount of bending, or refraction, is not the same for all colors of light. This means that a simple lens has different focal lengths in red, yellow, green, and blue light; the image will be a series of overlapping images in different colors. This error is chromatic aberration or primary spectrum; it makes simple lenses difficult to use as telescopes. However, by combining two or more lenses made of different types of glass, it is possible to construct an achromatic lens that brings two or more colors to the same focus, making refractors practical. The theory behind this is explained in Chapter 2.

REFRACTOR VS. REFLECTOR

Given the color errors inherent in the refractor, why would anyone choose one? A good 6″ achromatic doublet shows about the same detail in lunar and planetary images as a typical 8″ reflector does. The main reasons are: (1) optical fabrication errors are less critical for lenses; (2) the optical surfaces are smoother; (3) its closed tube eliminates tube currents; and (4) the unobstructed aperture procures a cleaner Airy disk, thereby giving more contrast.

Building a 6″ f/15 Refractor

8

The main points in favor of reflectors are: (1) they are totally color-free and (2) far less expensive than refractors.

How detrimental is the chromatic aberration? At high magnification, stars show spurious color effects, but these are visible only around bright objects. The diffraction pattern of a fourth or fifth magnitude star generally looks textbook perfect. For 6″ f/15 and smaller doublet refractors, chromatic aberration is annoying, but not a serious problem.

An apochromat triplet has secondary spectrum about 1⁄10 that of a doublet, so for all intents and purposes, color is entirely absent. Apochromats cost about twice what achromats cost, so what you might spend on a 6″ apochromatic triplet would purchase a finished 12″ mirror or a 6″ apochromatic doublet plus a 10″ mirror. On the other hand, aperture for aperture, nothing beats an apochromat.

Refractors also have the problem of being difficult to mount. Because they are long instruments, refractors tend to oscillate easily. The long tube possesses considerable vibrational energy that the mount must absorb if the telescope is to be steady. The only recourse is a solid, well-damped mounting. Although an equatorial mounting may be preferable in theory,

A refractor is much longer, heavier, and taller than Newtonian reflectors of the same aperture.

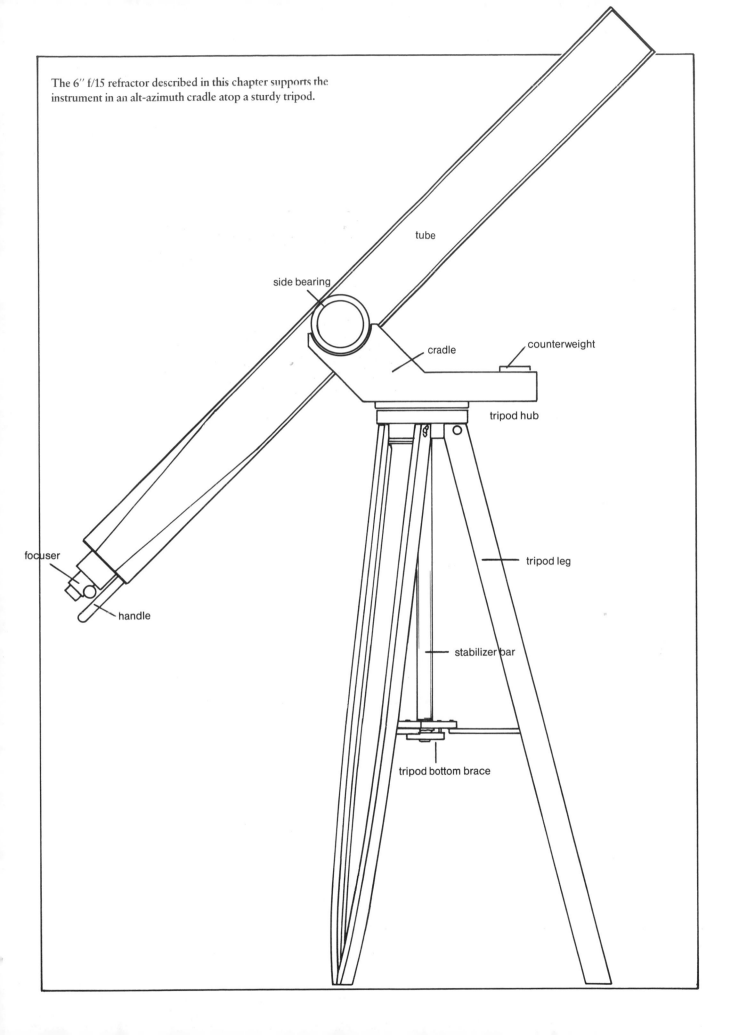

The 6″ f/15 refractor described in this chapter supports the instrument in an alt-azimuth cradle atop a sturdy tripod.

tube

side bearing

cradle

counterweight

tripod hub

focuser

handle

tripod leg

stabilizer bar

tripod bottom brace

the alt-azimuth cradle described here offers better vibration damping and greater overall utility than most home-built equatorials.

In the seventeenth and eighteenth centuries, refractors often had light, wooden tubes—usually built like a barrel, from thin staves. Nineteenth-century refractors had brass tubes, but brass is expensive and heavy. Plastic tubing (such as irrigation pipe) and spiral-wound paper tubes, while rigid enough for a stubby reflector, are simply not strong enough for the diameters and lengths needed in a 6″ refractor. One solution is to mount the lens in a light, thin-walled aluminum tube. Another, older solution, adapted to modern methods for this book, is a light wooden tube.

PLANNING

In terms of the total input, the 6″ f/15 refractor is by far the biggest, most time-consuming, and most expensive project in this book. However, it's an outstanding instrument that gives impressive, satisfying views of the moon and planets, of double stars, and deep-sky objects.

If you buy the objective lens (as, in fact, most people who make refractors do), buy a *mounted* objective. Mounting the objective yourself is a troublesome job and requires a machine shop in order to be done right. If you want a hand in every phase of construction, grind and polish an objective lens. (At least one firm listed in Appendix B sells glass and abrasive kits for 3″, 4″, and 6″ refractors; Appendix C gives references on grinding telescope objectives.) Be warned that making a lens is a *big* job.

Modify the mounting only in the direction of greater stability. Do not attempt to lighten it or develop a more portable version—a refractor needs all the stability it can get. Feel free to beef up the tripod and alt-azimuth head for a nontransportable instrument.

To build the refractor to these plans exactly, you'll need the following:
- one sheet of ¼″ A—C exterior plywood
- one sheet (4′x8′) of ¾″ A—C exterior plywood
- three 1″x6″x6′ of board lumber, your choice of species (pine works, but oak is better)
- six 1″x4″x6′ of board lumber, same species as the previous item

- 26 linear feet of ¾″x¾″ molding
- 36 linear feet of ½″ quarter-round molding
- one sink cutout at least 12″ square

For best result, buy the wood for this project several months before starting construction and allow it to come to equilibrium with the air.

It's difficult to estimate the cost of all the miscellaneous hardware, pipe, primer, and paint; but assume it won't total under $100 and will be spent $10 here and $20 there. Obtain a 30″ length of 1½″ iron pipe, threaded at both ends, a 1½″ floor flange, and a 1½″ to 1″ reducer—all from a plumbing-supply store. Start looking for 30 pounds of inexpensive counterweight material right away; try a local scrap-metal dealer for a 30-pound chunk of lead or cast iron, or contact a house wrecker for old lead plumbing. For the tailpiece, obtain a metal or Bakelite tube 4″ in diameter and 12″ long, a focuser ($60 if purchased), a star diagonal ($50 to $150), several eyepieces (at $25 to $75 each), and a finder telescope (see Chapter 9).

It doesn't matter whether you make the refractor tube or the mounting first, but it can be pretty frustrating to finish the tube only to find yourself more or less unable, because of the length of the tube, to test it *until* the mounting is built. Probably the best course is to build them together or plan for both to reach completion at about the same time.

If you make the objective yourself, the telescope should be near completion by the time to test the objective. A refractor objective is almost impossible to test on stars without a tube and mounting. One of the beauties of making a lens is that as soon as the glass polishes out, it's ready to use. With a small reflector, you really can't see much until the mirror has been aluminized.

CUTTING PARTS

Begin by cutting the long sides for the tube from a sheet of ¼″ plywood (grade A-C exterior fir or better, depending on how you intend to finish it). Each side is 7½″ wide and 96″ long. The easiest way to cut them is with a power saw, using a fine-tooth plywood blade. But be careful: Power saws are dangerous. Place four 2″x4″'s on the floor and place the plywood, best side down, on them. Mark the first cut. To the

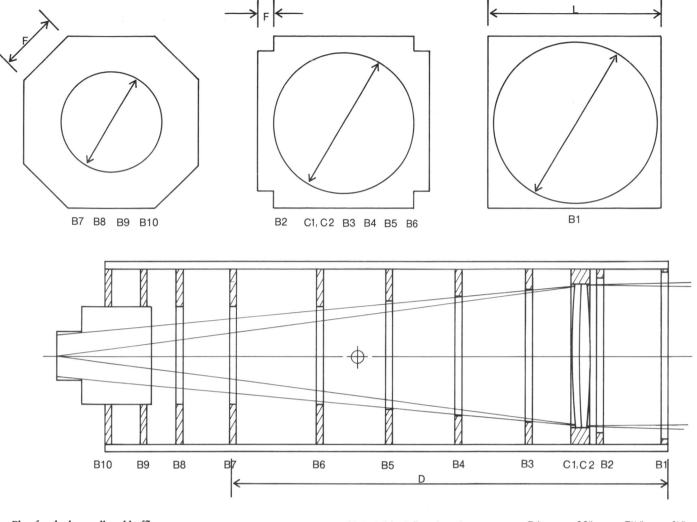

Plan for the lens cell and baffle parts.

#	D	L	F				
				B4	36"	7½"	¾"
				B5	48"	7½"	¾"
B1	0"	7½"	0"	B6	60"	7½"	¾"
B2	12"	7½"	¾"	B7	75"	7½"	½"
C1	–	7⅜"	⅞"	B8	84"	7½"	1½"
C2	–	7⅜"	⅞"	B9	90"	7½"	2¼"
B3	24"	7½"	¾"	B10	96"	7½"	3"

Note: table defines lengths

plywood, clamp an 8′-long board with a straight side to serve as the saw guide. Determine the offset between the guide and the blade, set the blade to the proper depth of cut (one full tooth should emerge from the opposite side of the wood), double-check all dimensions, then cut off the first 7½″-wide strip.

If you're using a hand saw, mark the plywood carefully. Support the plywood sheet on sawhorses or a low worktable so it cannot sag. Clamp it, and concentrate on keeping the cut straight and the saw blade perpendicular to the wood. A sharp, fine-toothed saw cuts ¼″ plywood quite quickly, no more than five minutes of sawing per 8′ length.

Next, make the twelve 7½″ squares of ¾″ plywood. These are baffles B1 through B10 plus the lens cell parts, C1 and C2, so they should match each other

closely in size. Saw two 7½″-wide strips from the edge of the sheet, then cut them into 7½″ lengths. After cutting, select two that are slightly smaller than the rest; mark them as pieces C1 and C2 for the lens cell and set them aside.

In each of the ten baffles, cut the diameter as shown in the plans. Draw lines with a straightedge or ruler from corner to corner to locate the center; then scribe the radius with a compass and cut it with a keyhole or saber saw. This is a tedious job, so intersperse it with other work. Clean cuts and exact diameters are important in only four of the baffles: B1, B2, B9, and B10. Baffle B2 holds the lens cell, and B1 is the front end of the tube; these cuts should be smooth and accurately round for cosmetic reasons. B9 and B10, at the back end of the tube, hold, and therefore

The ten tube baffles after having been cut.

must fit, the tailpiece tube tightly. Make sure that these holes are centered accurately; the alignment of the tailpiece tube depends on their accuracy. Cut them slightly undersize, then file the holes out to their full dimension.

The four corner strips are ¾″ square lengths of molding material selected for straightness. Begin by cutting all four of them 75″ long. Cut notches ¹⁄₁₆″ oversize in each of the first six baffles, B1 through B6, so they match the profile of the corner pieces with a loose fit, and make the diagonal corner cuts on baffles B7 through B10 for the octagonal end of the tube.

The mounting consists of the cradle, the tripod hub, and the three tripod legs. While the cradle and hub are made from plywood, the tripod legs are made from board lumber. This can be a soft wood such as pine, which is easy to work, or a hard wood such as oak or maple. In either case, the legs require six 1″x4″s,

Plan of the mounting cradle parts.

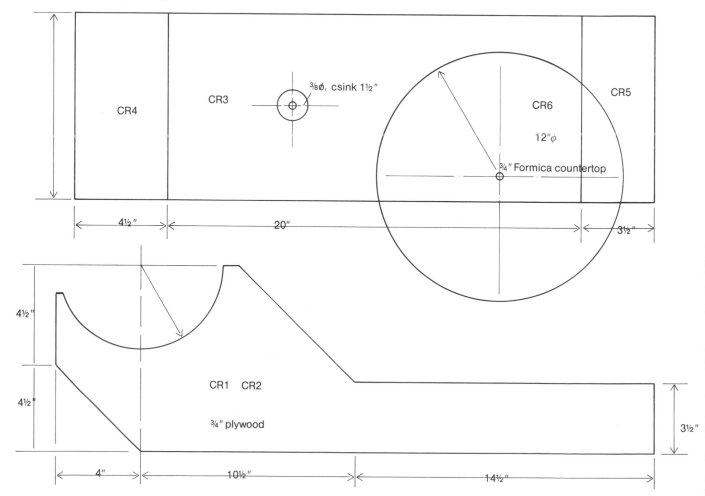

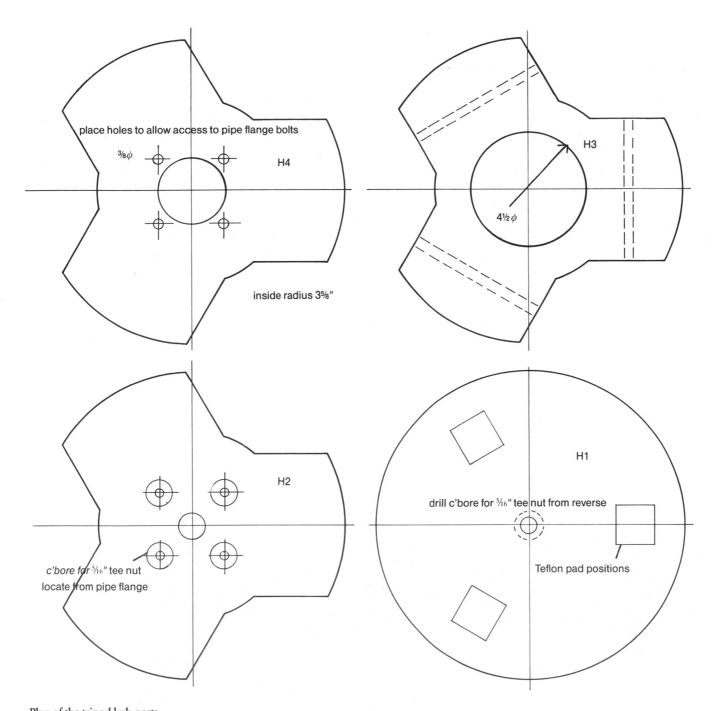

place holes to allow access to pipe flange bolts

⅜φ

H4

inside radius 3⅝″

H3

4½φ

H2

c'bore for ⁵⁄₁₆″ tee nut
locate from pipe flange

H1

drill c'bore for ⁵⁄₁₆″ tee nut from reverse

Teflon pad positions

Plan of the tripod hub parts.

three 1″x6″s, and about 32′ of trim—½″ quarter-round looks nice. At this stage, you don't have to do anything more to them.

Begin by drawing the irregularly shaped cradle side boards, CR1 and CR2, on the remaining ¾″ plywood. Also lay out the three trefoil layers of the hub, H2, H3, and H4; the round top piece of the hub, H1; the

cradle base, CR3; the cradle front (CR4) and back (CR5) boards. Since the trefoil pieces must match each other closely, draw a template on stiff cardboard, cut it out, then trace around it with a pencil. Make all the cuts carefully. After cutting, check to see that the pieces actually line up; number them and mark their fit with a pencil. Drill a 2½″-diameter hole in the

Cutting the three trefoil hub pieces with a saber saw.

As you assemble it, make sure the tube remains square.

bottom piece, H4; a 4″-diameter hole in the middle piece, H3; and a 1½″ hole in the top piece, H2. These holes accommodate the pipe flange that holds the stabilizing pipe.

Nail the cradle side boards, CR1 and CR2, lightly together in the best-fitting position. Plane, rasp, or surform them to the same dimensions. Pull them apart, switch sides, renail them, and again work them. This technique ensures a close-shaped match. Check to see that the cradle base, front, and back boards fit and have exactly the same width. If they don't, work them until they do. The final piece of the cradle is a 12″-diameter Formica (or other hard-plastic laminate) circle, C6, cut from a sink cutout. The surface of this piece is the azimuth bearing that the telescope turns on.

Your telescope is now a rather large pile of cut lumber. There are more pieces to make, but since they must be cut to fit, wait until their actual dimensions can be determined from the partially finished telescope.

ASSEMBLING THE TUBE

The tube is a tricky structure to assemble. The strategy is to nail and glue the corner strips to the two sides, then tack the third side in place, set the baffles approximately in place, square up all the parts, then glue and nail all the baffles (except B2, which supports the lens cell) in place. After the glue dries, you'll glue and nail the third side on. When assembled with three sides, the tube is rigid and quite strong. You will then test the telescope before painting it or putting on the fourth side.

Begin with the corner strips. Attach two to each of the two ¼″ sides. Make all the joints with water-resistant wood glue (Elmer's Carpenter's Glue or Panite Plastic Resin Glue are excellent) and small finishing nails or ¾″ brads spaced 2″ apart.

Next comes a tricky part: putting the baffles between the sides. It's extremely helpful to have someone else hold the sides while you apply the glue and nail the first few baffles. The easiest order is B1, B7, B3, B6, B4, B5, and B8. Do *not* glue B2 now.

Slide the tailpiece through B9 and B10, then clamp them lightly in place at the end of the tube. Sight down the tailpiece tube to make sure that it's aligned

Baffles B9 and B10 must be placed so the tailpiece is aligned with the axis of the tube.

The tailpiece tube shown in place.

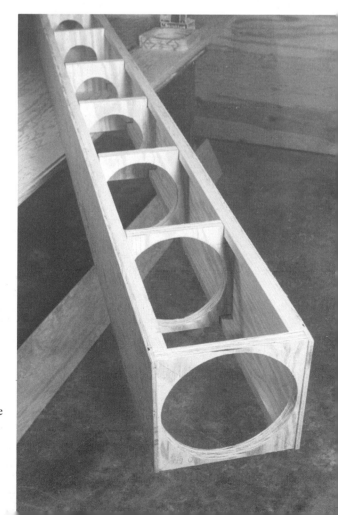

The front end of the tube; the objective lens is attached to the second baffle, B2.

with the axis of the tube. If it is angled, slide B9 or B10 slightly up or down, rotate one around the tube, or, if necessary, sand the edges of one or both to attain alignment. Glue and nail these baffles in place. The last step—for now, at least—is gluing and nailing the third side on. Once again use plenty of small brads. Allow several days for the assembly to dry.

The balance point for the tube determines where the side bearings should be, but these are difficult to install once the tube is closed. The easiest route is to attach them to the partially completed tube and then balance the tube with small weights after completion. Determine where the tube balances by dummying all the remaining parts together (i.e., the fourth side, the tailpiece, the objective and cell, the star diagonal, the eyepiece, and the finder telescope) and then finding the "teeter point" with a dowel. Attach the side bearings with stove bolts, engaging tee-nut fasteners. Use a carpenter's square to make sure the side bearings lie exactly opposite each other.

Do *not* complete the tube at this time. Wait until the lens and its cell are complete, until you have made sure that everything works and balances properly in the mounting, and until the tube interior has been sealed and painted black.

ASSEMBLING THE MOUNT

The cradle is a straightforward box with two unusually shaped sides. Assemble it first with 4-penny nails to check that the sides lie opposite each other; then knock it apart, apply glue to all the joints, slip the nails back into their holes, and put the box together. After reinforcing the assembly with #10x 1½" wood screws, nail and glue the Formica-covered circle, CR6, to the bottom. After the glue has set, round the corners and edges, then sand them down. Drill the center hole. If you wish to, round the corners of the telescope cradle with wood filler.

Cross section of the mounting cradle.

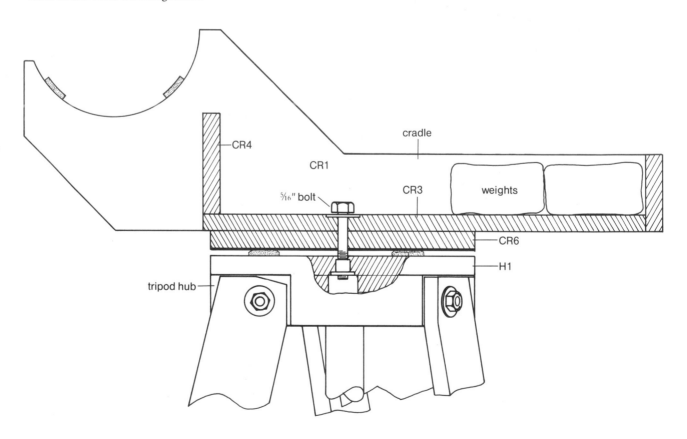

Next, assemble the hub. This massive block of wood weighs about ten pounds when finished. Bolt the 1½″ pipe flange to the top trefoil piece, H2, using ¼″-20x1″ flat-head bolts into tee nuts. Glue and nail all three trefoil layers, H2, H3, and H4, together, and clamp them. Be extremely careful to get the layers squarely on top of one another. Check this with a square as you work; otherwise the legs won't bolt on squarcly. When the assembly is dry, rasp or surform the sides exactly square and the outside curve as smooth as you can get it.

Drill the hole for the center bolt and insert a tee nut from the reverse into H1, the fourth hub lamination; then glue H1 to the hub assembly. When this is dry, fill out irregularities and voids to improve the hub's cosmetic appearance; then rasp or surform and sand it smooth.

The most critical operation in the hub is the drilling of three ⅜″-diameter holes through the bosses on the hub. These holes hold the tripod legs, so they

Bottom view of the cradle, showing the Formica bearing surface.

Cross section of the tripod hub.

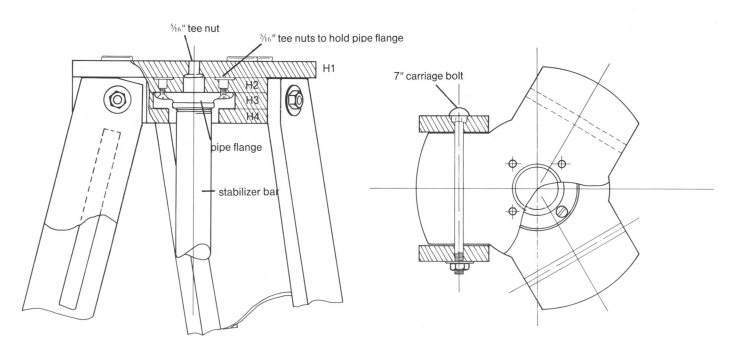

The partially completed tripod hub with one leg attached.

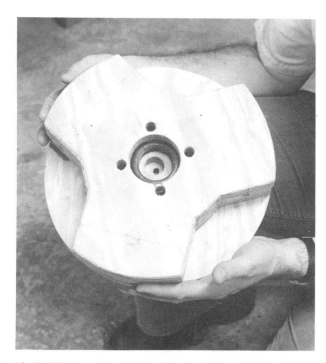

The hub from below; four small holes allow tightening the bolts that hold the pipe flange to the hub.

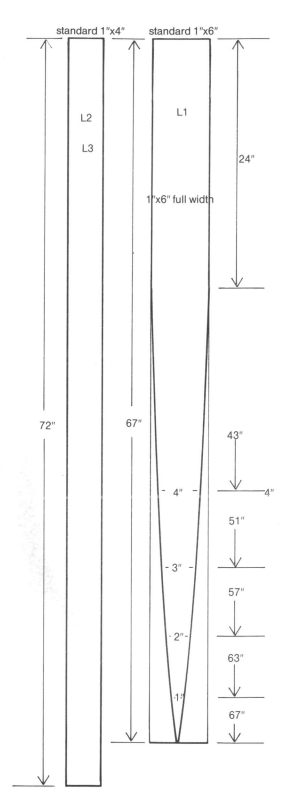

Plan for one tripod leg; make three.

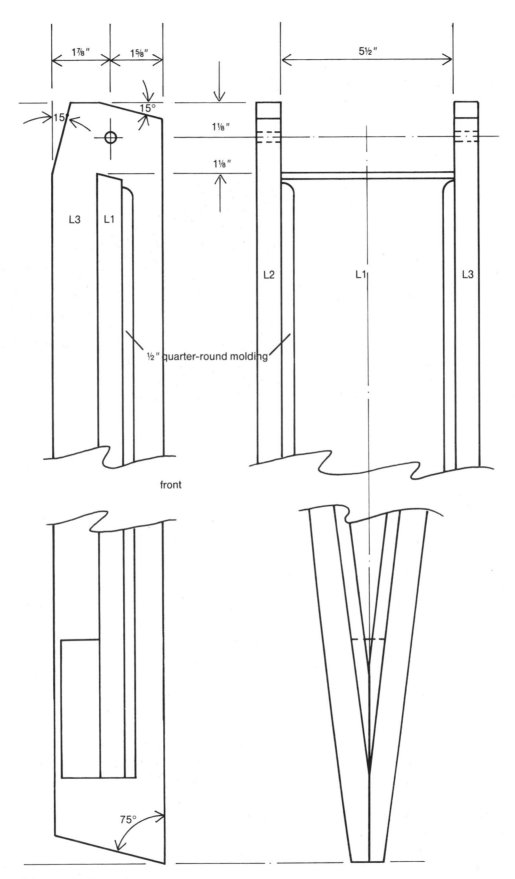

front

A cross-section view of the assembled tripod leg.

must be perpendicular to the center of the hub and parallel to its top surface. Mark and drill them as accurately as you can. If your drill is not long enough, drill in from both ends: If the holes meet in the middle, you'll know they're not too far off. After drilling them, file them out smooth and straight.

The legs are the most involved part of the mounting. Throughout their construction, keep checking the parts against each other—they must match. Begin by cutting the 1″x6″s (1L1, 2L1, and 3L1) to 67″ in length, and the 1″x4″s (1L2, 1L3, 2L2, 2L3, 3L2, and 3L3) to a length of 72″. Cutting the curves in the 1″x6″s is next. This requires determining the natural curve the 1″x4″s assume when bent. Tack-nail two cutoff ends of a 1″x6″ to a pair of the 1″x4″s (an L2 and L3) on top of a 1″x6″ L1, so the ends protrude about ¼″ beyond its end. Clamp the small pieces to it. Now you're ready to bend the free ends of L2 and L3 together until they touch. They should just meet at the end of L1.

Plane the material from the inside ends until the tip is the thickness of one board. Clamp them together, then trace the smooth curve of the inside edges of the 1″x4″s onto the 1″x6″, L1. Disassemble them, saw off the material outside this curve; then plane, rasp, or surform it to match the smooth, penciled curve. After checking to see that L2 and L3 still bend smoothly along the edges of L1, trace the curve onto the other two L1 pieces; then cut and plane them to the same shape. Mark each part as belonging to a particular leg, and check to see that the width of each L1 piece matches its boss on the hub.

Mark the 1″x4″s 2¼″ from the top and roughly 2½″ from the bottom. Align the upper mark on L2 and L3 on the side of each L1. Clamp their upper ends. Working on a flat surface, place blocks of 2″x4″ to hold the profiled L1 piece at the correct height, bend L2 and L3 over L1 again, mark any remaining high spots on L1, disassemble the leg and plane or rasp or surform them off, then reassemble the leg and check it again. When L2 and L3 fit face-to-face, clamp them in place and tack the leg together lightly with a dozen 4d finishing nails. Assemble each of the legs in this way.

Next, on the outside faces of L2 and L3, drill and countersink pilot holes for #12x2″ flat-head wood screws 4″, 14″, 24″, 34″, 44″, 54″, and 64″ from the top of each side. These are 1⅝″ from the front side of the leg and enter the mid-line of L1. Also drill ⅜″-diameter holes 1⅛″ from the tops of L2 and L3 for the bolts that attach them to the hub. Disassemble each leg by partially drawing the nails out, remove all wood chips, and bevel the edges of each L2 and L3 and the top of L1. Then coat the edges of each L1 with glue, reassemble the legs again by putting the nails back in their holes, firmly clamp them, and run wood screws in tight, starting from the top.

When the glue is dry, glue in a small reinforcing triangle of wood to the lower end of the back side of each leg and glue and nail ½″ molding in the corners of the front side. After these are dry, shape the upper and lower ends of each leg with a 15° cutout, as shown on the plans. These cutouts permit the leg to bend outward from the hub or to fold in for carrying and storage.

If all has gone well, when you pound the ⅜″-diameter by 7″-long carriage bolt through the legs and hub, the legs will fit the hub tightly and will bend out at a 15° angle but still fold in without binding. The legs should lie equidistant from the hub pipe, should rest level, and should come out spaced equally on a 48″-diameter circle drawn on the floor.

This is the best time to rectify small construction errors. If the legs fit the hub loosely, buy 2″- or 2½″-diameter fender washers to fit between the legs and the hub. If the legs are too small to fit over the hub, remove wood with a plane and rasp or surform from the inside of the leg until it fits. If one or more legs is not straight, the problem probably lies in the straightness of the hole through the hub. Using a large file, carefully enlarge one end of the hole to bring the leg into line. When the leg fits correctly, it should be possible to stop its swing motion by tightening the wing nut on the bolt, and the clamped leg and hub should be almost as rigid as a single piece. Any looseness in these joints shows up as image vibration and lost motion in pointing the telescope.

When the tripod passes all the tests, cut the parts for the bottom brace, altering the dimensions, if necessary, to ensure that they fit. Install them and their hinges. The bottom brace stops unwanted motion. When you install it, you'll be pleased—the whole tripod will become amazingly rigid.

Nailing the upper end of one side of a leg into place just prior to attaching it with wood screws.

Clamping the bottom end of a leg in order to glue and screw it together.

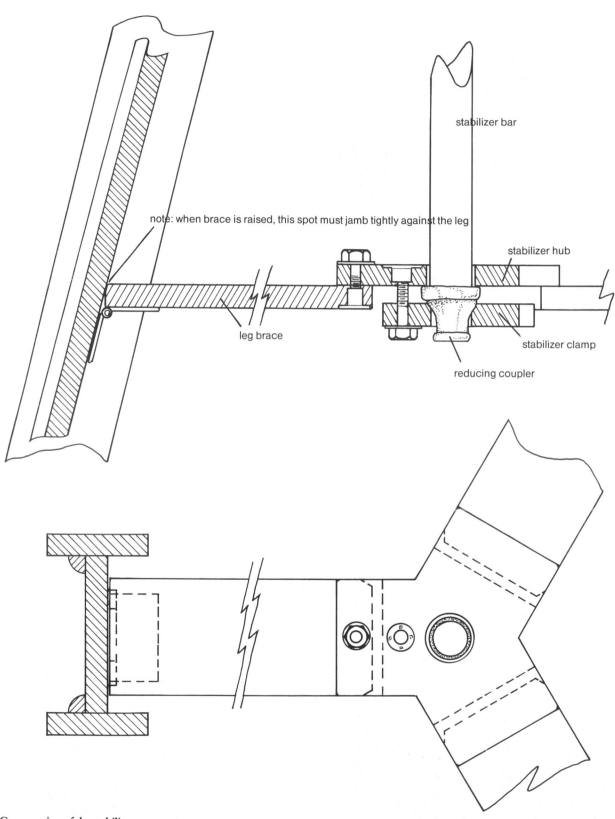

note: when brace is raised, this spot must jamb tightly against the leg

stabilizer bar

stabilizer hub

leg brace

stabilizer clamp

reducing coupler

Cross section of the stabilizer.

The completed stabilizer installed in the tripod.

MOUNTING THE OBJECTIVE

If you plan to purchase the objective, it is best to buy one already mounted in a "lens cell," that is, a metal holder that keeps the lens elements accurately positioned relative to one another. A ready-made cell saves the effort and expense of making a part that often requires turning on a lathe to close tolerances, and it does away with the dangers inherent to handling the objective lens.

Turn-of-the-century lens cells were usually made of brass with a flange, or collar, containing three sets of push-pull bolts for "squaring-on" the objective; they were typically 1″ to 2″ larger in outside diameter than the aperture of the objective. The lenses themselves were secured in the cell by the narrow lip of the cell and spaced with small pieces of tinfoil or celluloid. Disassembling an old objective from its cell for cleaning or refurbishing is a tricky job. You run the risk of severely degrading the lens's performance if it is not reassembled in the cell exactly as it originally was. Because nineteenth-century lenses were not always well-edged or exactly centered, they were sometimes shimmed in odd positions to

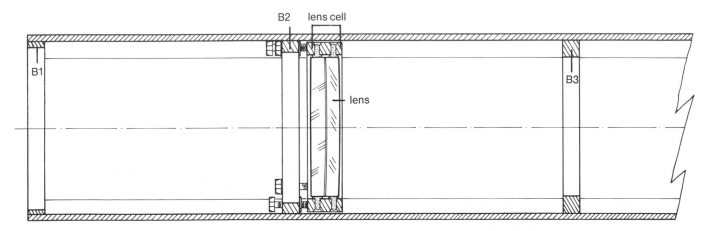

The upper end of the telescope tube, showing how the objective cell is positioned in the tube.

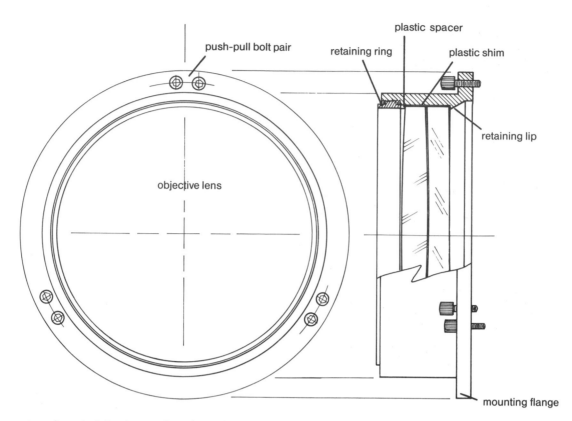

Cross section of a typical aluminum or brass lens cell.

compensate for these fabrication errors.

Since many of today's commercially made lenses are designed for use with thin-walled aluminum tubing that the lensmaker also sells, the lens cells are designed to fit right into the tubing with no provision for alignment other than the (probably justified) assumption that the end of the tubing is cut off square. The cell is usually a simple piece of tube with a wall thickness of ¼″ or less, and the lens fits in it snugly. While modern lenses are edged and centered so they fit these mass-produced cells quite reproducibly, it is nevertheless unwise to disassemble a lens unless absolutely necessary.

INSTALLING A MOUNTED OBJECTIVE

Installing a mounted lens in the wooden telescope tube is a simple job. Cut a hole about ⅛″ larger in diameter than the outside of the lens cell in the two remaining 7½″ squares of ¾″ plywood, C1 and C2.

Glue the pieces together; rasp, surform, and sand them to fit with roughly 1/32″ clearance into the upper end of the tube. Note that two sides of the wooden cell holder are beveled to allow removing the cell from the tube. After beveling, test to see that you can insert and remove it.

Clamp the cell to baffle B2. This baffle holds the cell holder. Drill three ¾″-diameter holes ⅛″ deep in B2, then three concentric 5/16″ holes through both B2 and the cell holder for the three pull bolts in the corners of the cell and B2 at the location shown in the plans, and three 5/16″ holes through B2 for the push bolts. Screw three small mending plates where the push bolts will press against the cell. Insert tee nuts and the 1¼″- and 2¼″-long ¼″-20 hex-head bolts; then check to see that the assembly works and fits the tube.

At eight points around the cell, drill ¼″-diameter glue injection holes and carefully chisel out a glue channel on the inside diameter. Thoroughly clean all chips from the holes, and then give the entire cell a thorough brushing with a coarse wire brush to remove any weak or damaged wood fibers.

Seal the cell and baffle B2 with penetrating sealer; then fill dents and holes with wood filler, reseal, allow it to dry, and, finally, paint it. The cell and its supporting baffle must be particularly moisture-resistant because changes in them can affect the alignment of the objective.

When the paint has completely hardened (allow at least two weeks), glue the lens cell into the wooden cell holder by injecting silicone sealer (or "aquarium cement") into the glue injection holes. This material remains soft and resilient for years and cushions the lens from shocks. Begin with placing the cell and holder on a flat, clean surface. Make absolutely sure that the front side of the objective and the front side of the cell holder face in the same direction. Space the cell accurately in the center of the hole with toothpicks or bits of cardboard, then inject glue into the holes. Do not allow any excess to flow out near the optical surfaces of the lens. Allow the glue to set for at least 24 hours before moving anything, and then pick it up cautiously.

If it ever becomes necessary to remove the cell, you can cut the glue with a narrow-blade knife or a wire loop without any damage.

INSTALLING AN UNMOUNTED OBJECTIVE CONVENTIONALLY

If you have made the objective or purchased one that is not mounted, you must now mount it. If possible, machine a conventional cell for it from aluminum or a clean-machining material such as Bakelite. The elements are usually spaced by a thin ring about 1/16″ wide. Its thickness defines the spacing. Alternatively, if the spacing is small, three small tabs of soft metal foil held in place with a trace of acetone solvent glue can serve as separators.

The inside dimension of the cell should be about .010″ greater than the outside diameter of the objective, and the gap should be filled with a resilient spacing material such as thin plastic shim stock. Although the lens is secured by rings machined into or screwed to the faces of the cylindrical body of the cell, rubber

Cross section of a plywood-and-glue lens cell.

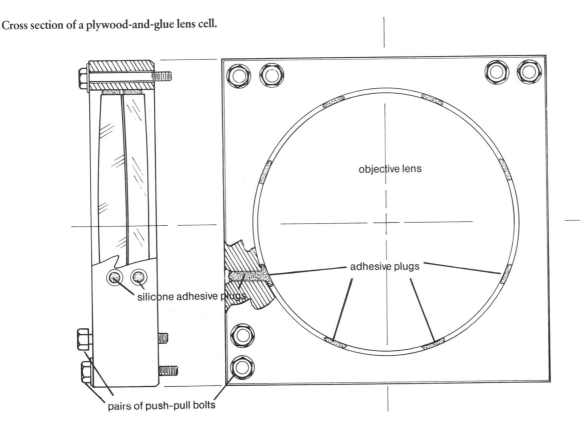

silicone adhesive plugs

pairs of push-pull bolts

objective lens

adhesive plugs

O rings with ¹⁄₁₆″ to ¹⁄₃₂″ cross section are all that actually touch the glass. The pressure on the lens should be just adequate to prevent the lens elements from rattling when you shake the cell.

INSTALLING AN UNMOUNTED OBJECTIVE UNCONVENTIONALLY

If you do not have access to a machine shop or cannot find someone to machine a holder for you, it is possible to tape an objective and mount it directly in the holder. The prime requirement in mounting an objective is that the lens elements remain accurately concentric; stiff Mylar or metal-backed tape can do the job. As a precaution, inspect the assembly every few months for degradation, and monitor the image quality for "creeping" of the lens elements.

Cut the inside diameter of the cell holder about ¹⁄₈″ larger than the outside diameter of the lens, but otherwise finish it as you would for holding a mounted lens. When you are ready to assemble the lens, set up a clean work area covered with freshly washed old linen or a clean cotton cloth. Place the flint element back-side-down on a low cylinder covered by soft lens tissue.

Using the merest touch of acetone solvent glue, attach three soft metal foil spacers of the correct thickness, intruding about ¹⁄₈″ onto the finished surface to the rim of the flint glass element. With a clean plastic squeeze bottle or rubber syringe, blow off any dust particles, then lower the crown element onto the spacers. Center it by feel—your fingertips can detect an eccentricity of .001″ if the elements are of the same diameter and accurately round. Stick three pieces of tape 1″ wide at equal intervals around the outside of the objective to secure the lens temporarily.

Cut a band of tape just wide enough to cover the gap between the elements; wrap it around the entire periphery of the lens. Lower the cell holder over the assembled lens and shim it around the rim so the lens lies at its center. Center it with shims (toothpicks), then inject aquarium cement through the holes. Take care that no cement spreads to the optical surfaces of the glass. Allow 24 hours before installing the objective in the telescope.

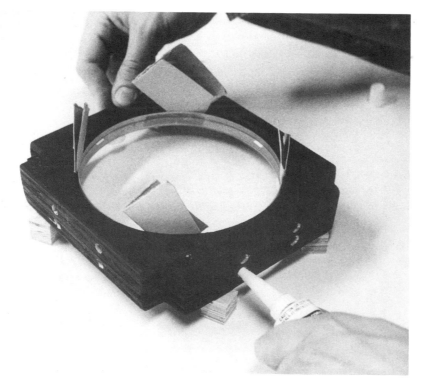

Injecting silicone adhesive between the lens cell and the taped-up objective.

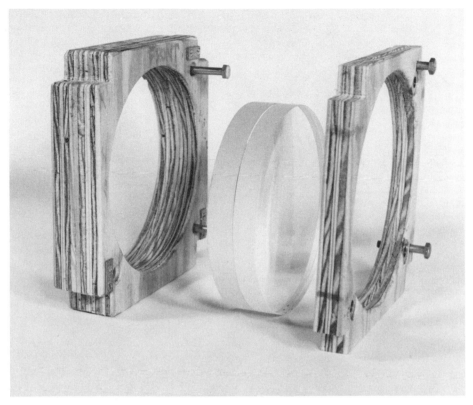

Baffle B2, the lens cell, and the lens elements.

TRYING IT ALL TOGETHER

You are now almost ready to try the telescope for the first time. Cut out and screw in place four Teflon bearings for the cradle and three for the top of the tripod. Use very small wood screws; #4x½″ screws work well. Nail baffle B2 in place at its nominal position with four small brads; then mount the objective cell on it, using the push-pull bolts.

At dusk, set up the tripod, put the cradle on it, and place the counterweights in it. Lift the tube gently and lower it onto the trunnions in the cradle. Balance the tube with small weights—C-clamps are particularly handy. The tube will suddenly seem very light. A gentle touch moves it.

Slip the star diagonal and the lowest-power eyepiece into the focuser, point the tube at a bright star or planet, and try to get a sharp image. If all is well, the object should come to focus in the middle of the focuser's range. If it is at one extreme or the other, slide the tailpiece to bring the focus to the middle

of the focuser's range. If it cannot provide enough adjustment, disassemble the telescope, move B2 in the appropriate direction, then reassemble it and try again. The move should not exceed an inch or two.

Once the focal distance is established, collimate the objective. A quick method is to shine a penlight into the focuser tube and look beside it. You will see three reflections spread across the lens. Adjust the push-pull bolt pairs until the images merge into one.

Now perform a quick check of the objective's image quality. Starting with inside focus, slowly turn the focus knob and run through focus. Check first to see that the collimation is good enough. Coma-corrected refractor objectives are surprisingly insensitive to alignment, so you may not need to do any more collimation at all. However, if the image appears oval or elliptical rather than circular, the objective is one that is sensitive to alignment. Swing the tube down, give one of the push-pull bolt pairs a small fraction of a turn, then test again. If the image becomes worse, turn the same push-pull pair in the other direction. If

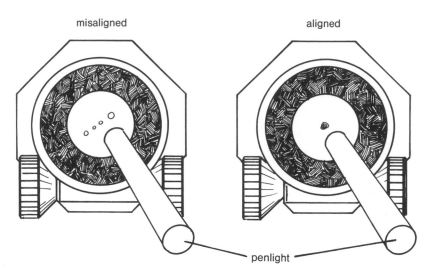

misaligned

aligned

penlight

Aligned and mis aligned objective lenses as tested by shining a penlight down the focuser tube.

The lower end of the tube has an octagonal cross section.

the image improves, turn farther in the same direction, or try adjusting the other push-pull pair. After getting good images at low magnification, switch to high magnification and finish the job.

Spend a few hours in getting used to the telescope. Does the mount perform properly, or does shakiness make viewing difficult? If it does, something is wrong; the tripod should be as solid as a rock. Is the amount of friction in the mounting right? Does the mount move easily enough but not too easily? Move the Teflon bearings farther apart or closer together to control the friction. Try splitting a few close double stars to see whether the objective performs to specs. If you've used reflectors but never used a refractor before, the secondary spectrum in the image of a bright star may distress you. On fainter stars, however, this color is not visible and should not prevent resolving right to the diffraction limit.

Having determined the position of baffle B2 and roughly assessed the quality of the objective and the performance of the mounting, you're ready to finish the mechanical parts.

COMPLETING THE TUBE AND MOUNT

After taking the telescope apart again, carefully saw away the excess plywood in the octagonal section at a 45° angle about ⅜″ above the diagonals on the tapers. Cut and fit two of the four triangular filler pieces for the third side by trial and error. Start with ¼″ plywood pieces roughly 22″ long, 3½″ wide at the large end, and ¾″ wide at the small end. Plane, rasp, or surform the edges to 45° so they fit the sides tightly, then glue them in place. For the other two, shape one side of the filler piece to fit against the side, then glue it in place. When the glue is dry, rasp or surform the top flat. The fourth side, which is still not attached, provides access to the objective and for cleaning. Cut a port in the middle, then cut off the upper 24″ of it. Drill and countersink holes for #8x¾″ flat-head wood screws in this piece, and attach it. Tack the side in place, and cut and rasp or surform the edges to match the octagonal section. Bevel all the corners on the tube exterior, then

The tailpiece tube holds the focuser. (right)

The refractor breaks down into three major parts; note that the tripod legs hook together for ease of transport. (below)

remove the fourth side in preparation for painting.

Seal the interior and exterior of the tube and both sides of the fourth side, allow them to dry fully, then paint all the interior surfaces flat black. Paint the inside of the tailpiece, then cut a baffle with a 2¼″ opening from thin cardboard and glue it to the inside end of the tailpiece. Paint it black too. Let everything dry thoroughly *before* gluing the fourth side on; paint dries poorly in closed spaces. Finally, paint the exterior any color you want. Use a tough, highly waterproof coating such as polyurethane floor-and-deck paint.

Disassemble the cradle and the tripod and remove the little pieces of Teflon. Correct any problems noticed in testing. Seal all the parts, allow them to dry fully, then paint them any color of your choice. Again, polyurethane floor-and-deck seems to be the toughest and most water-resistant material available in hardware stores.

FINAL ASSEMBLY

By now you should be pretty much an expert in assembling and disassembling your 6″ refractor. Put it all together again: Bolt the legs on the tripod, reattach the braces on the legs, and screw the Teflon bearings back in place. Slide the tailpiece into the end of the tube, bolt the objective back in place, attach a finder telescope (see Chapter 9), and put on the side bearings. Check the collimation, using the penlight method described earlier. Check to see that the assembled tube balances properly. Add weights to the front or back of the tube if there is a minor imbalance. Install carrying handles if you want them.

On a clear evening, once again set up the tripod, bolt on the cradle, lift the tube and set it on the bearings, and enjoy the feel as it swings around smoothly. Adjust the tailpiece for proper focus on the first star of the evening, and begin to observe.

To set up the tripod, swing the legs out, then raise the leg braces and bolt them into place.

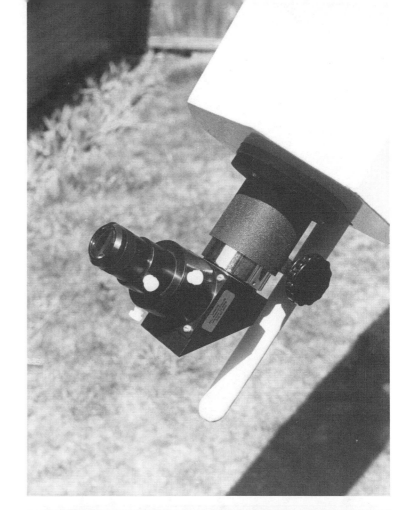

The finished telescope sports a finder on the side and a "steering bar" protruding from the end.

The tube's side bearings rest on Teflon pads in the cradle; the tube moves very smoothly under a light touch.

Cells, Spiders, Focusers, Finders, and Eyepieces

9

Telescope parts are available as standard, commercially manufactured, off-the-shelf units. You can buy them, bolt them on, and immediately begin observing. If you purchase all the necessary parts, you may find that their cost exceeds the outlay for the rest of the telescope, and, most unfortunately, the performance of some commercial parts leaves much to be desired though others are excellent. With a little ingenuity, you can often do better yourself and for a lot less money. The choice to buy or make remains yours. The sensible way is to make what you can and buy the rest.

MOUNTING THE PRIMARY

The primary mirror must be held firmly, gently, adjustably, and permanently. The first problem is: How gently, and how firmly? On the one hand, the mirrors are as thick as they are—usually ⅙ their diameter—in order to prevent them from flexing more than ⅛-wave of light *under their own weight*. If the forces we apply to hold the mirror in place exceed the weight of the mirror, the mirror will flex. On the other hand, coma exceeds ¼-wave 0.047″ from the optical axis of a 10″ f/6 mirror. The mirror must not move more than a tenth of that amount—a few thousandths of an inch—or optical performance will be adversely affected.

Furthermore, the cell that holds the mirror must be adjustable; a range of motion an ⅛″ across a width of 10″ is sufficient. And you will *hope*, at any rate, that once you've aligned it, the cell will remain aligned despite the considerable bumping and banging that accompanies normal use.

Mounting with Adhesives. Gluing the mirror to its support is not a new idea, but it has become practical only as satisfactory glues have become available. Strong and rigid adhesives (e.g., epoxies) work poorly because they transfer stresses from the cell to the mirror. Solvent-based contact cements, although flexible, work poorly because the bond must be thin. The cement cannot flex to relieve stress. Only if the mirror is glued to a flexible support, such as thick felt, and that glued to the cell, is it isolated from stress.

Silicone rubber adhesive, however, is nearly ideal for mounting mirrors under 12″ in aperture. It bonds

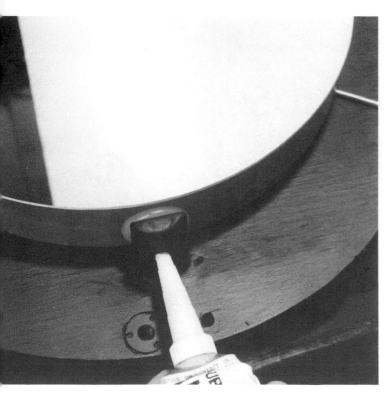

Silicone rubber adhesive bonds tightly, remains flexible, and is easily injected between the mirror and its cell.

tightly to glass, remains flexible for years, and can be injected between the mirror and the support to form a thick, flexible pad. Furthermore, since the cell becomes an integral part of the telescope, it is unnecessary to buy a separate mirror cell.

An adhesive-mounted mirror must be supported both longitudinally and radially—that is, from behind and from the sides. For small mirrors, three equally spaced pads $3/16''$ to $1/8''$ thick and $3/4''$ in diameter behind the mirror, plus three equally spaced pads around the rim, make an adequate support. As a precaution against its becoming unstuck, the adhesive must be "captured" between the cell and the mirror with an irregularly shaped "foot" on the back.

Flotation.   Even when supported without stress, large, thin mirrors flex under their own weight and require "flotation." To flex appreciably, the mirror must be larger than $12''$ or thinner than one-ninth its diameter.

Suppose we divide the mirror into nine zones of equal area: an inner zone supported at three points

and an outer zone with six points; and then into thirds circumferentially. Three support triangles placed 120° apart on a radius of 0.304 times the diameter of the mirror, with support points at the apices of an isosceles triangle with a base 0.408 diameters long and 0.353 diameters high, will float the mirror. The pivot point of each triangle lies on the shortest bisector 0.101 diameters from the inner support point.

Aluminum is excellent for the triangles; it's light, thin, and strong. For a $10''$ thin mirror, $1/8''$ aluminum is thick enough; for a $16''$ mirror, use $3/16''$ aluminum. Mark and drill holes for small round-head bolts on one side, and drill a 60° cone on the reverse side for the balance point. Balance the triangles on $1/4''$ acorn-nut heads. The triangles must be free to teeter and thus to distribute the weight, yet they must be constrained from falling off the acorn nuts. Drill small holes in the triangle plates and run thin metal pins through them to hold them in place relative to the cell.

On the whole, it is far easier to mount a mirror in an alt-azimuth telescope than in an equatorial because the mirror tips only in one plane. A low-cost alternative method of flotation employed in large Dobsonian telescopes employs plywood triangles held in position by flexible plastic sheeting and a thin metal or plastic "sling" cradling the mirror radially. The mirror rests on the supports, held only by gravity. These minimal cells work well for large apertures, but they are impractical for small mirrors. Because observers with large Dobsonian Newtonians may travel considerable distances to good skies, the mirror is removed when the telescope is not in use.

Metal Cells.   These are available from many of the suppliers listed in Appendix B. Unfortunately, such cells are not the best for mounting small mirrors because they usually do not meet the conditions for satisfactory mirror-mounting without careful adjustment.

The mirror typically rests in an aluminum framework that partially encloses it and is often cut away to allow ventilation or to reduce weight and material. The mirror is prevented from falling forward by clips extending $1/4''$ or so across its face; it is restricted laterally by cork or neoprene pads that are sometimes adjustable. In some designs, the cell's orientation is set with wing nuts opposed by springs on bolts

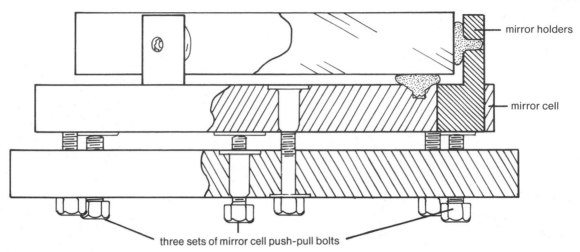

Cross section of an adhesive-bonded mirror cell.

Layout of the triangles for a nine-point flotation cell. All units are given in terms of D, the mirror diameter.

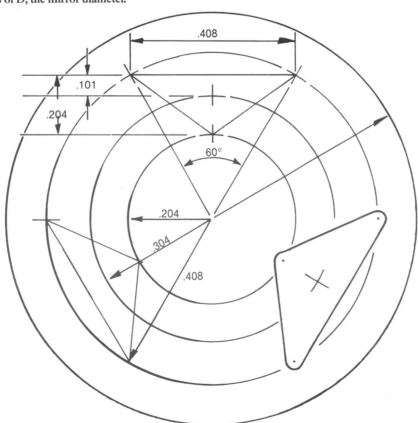

Mirror cell, with flotation, for a large mirror; triangles are
made of aluminum.

extending from the cell support. Since springs can
wiggle and wing nuts tend to vibrate loose, the cell
shifts and the mirror moves in its cell. In short,
the mirror must be realigned often.

Furthermore, these cells encourage clamping the
mirror down "good and tight," thereby severely
flexing it. No lasting harm is done. The mirror will
give good images as soon as the stress is relieved,
but until the problem is corrected, the telescope's
images may be terrible.

The easiest method for installing a mirror in a
metal cell is to place an index card between the
mirror and each clip or adjusting screw, bring them
up until they touch it lightly, and then remove the
cards. The mirror is then free to rattle a few thou-

sandths of an inch. While this means that the mirror
is not stressed, it also means that there is too much
movement to assure critical optical alignment. The
result is that whenever the telescope gets knocked
around, the observer must realign it for top-notch
performance.

Luckily, a little misalignment or flexure seldom
interferes with observing, at least at low magnification. Critical observers eventually fine-tune the
cell to clearances considerably smaller than the
recommended few thousandths of an inch, or they
mount the telescope permanently so it seldom gets
bumped or jounced around.

Metal cells are necessary for mounting a mirror in
an equatorial telescope larger than 12″ in aperture

because adhesives cannot handle the mirror's weight. A drop of silicone adhesive between the pad and the mirror does, however, prevent that annoying few thousandths of an inch of rattling without stressing the glass.

DIAGONAL MIRROR CELLS AND SPIDERS

The Newtonian secondary hangs in the middle of the tube, where it blocks part of the incoming beam of starlight. It is mounted in a mirror cell and held in place by a spider. A spider is a simple one- or many-legged support for the diagonal mirror and its cell. The spider must block as little light as possible.

Diagonal Mirror Cells. The easiest way to hold a small secondary mirror (i.e., under 2.5″ minor axis) is by gluing the mirror to a support with flexible

silicone adhesive. For larger secondaries, use adhesives plus safety clips so you don't worry about the adhesive failing and one expensive piece of glass dropping on another.

The cell can be rigidly attached to the hub of the spider if the spider is adjustable, or adjustably if the spider is rigid. Longitudinal adjustment, rotational adjustment, and angular adjustments are needed. A rod passing through the spider hub provides the first two; push-pull bolts on the back of the spider cell provide the third.

There are three general philosophies on how to support the diagonal mirror cell.

The first is simplicity itself: Mount the mirror on the end of a stiff metal rod and bend the rod to adjust the mirror. There is really no spider or cell in this scheme.

The second is easy to make but hard to adjust: Make the spider rigid and attach it firmly to the mirror

Minimal mirror cell for a large-aperture Dobsonian telescope.

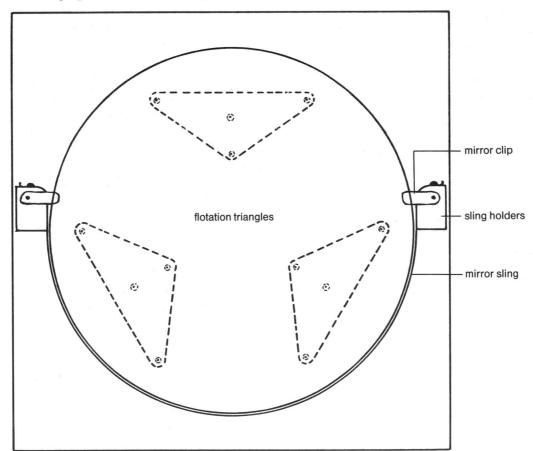

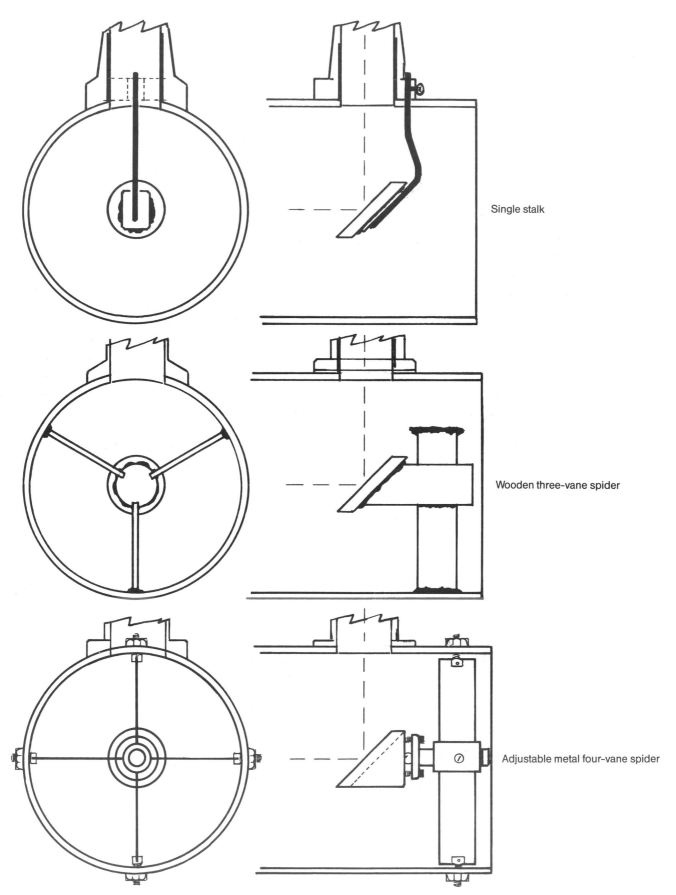

Single stalk

Wooden three-vane spider

Adjustable metal four-vane spider

Three styles of diagonal mirror mounts.

cell; then adjust the diagonal by sliding the spider legs in the tube.

The third is easy to adjust but hard to make: Thin spider legs are stretched tight across the inside of the tube, rigidly fixed to the tube. This supports an adjustable cell.

Applied intelligently, all three methods work well.

Single-Stalk Spiders. These work best for telescopes 4″ and 6″ in aperture. Their chief disadvantage is that if the stalk is too long or too thin, it will sag, bend, or vibrate. The stalk may extend from the eyepiece side, from the side opposite the eyepiece, or across the whole width of the tube.

The stalk is a brass rod roughly ⅛″ in diameter, with a brass pad to hold the mirror soldered or brazed to it. Bend the end of the rod into a ring and bolt it to the side of the tube, or insert its end, or ends, into a hole drilled in the tube. Some commercial focusers come with a stalk socket cast into the base. You might opt for a stalk of aluminum ⅛″ thick by 1″ wide rather than a rod. It will be less prone to vibration, and, seen edgeways, it blocks no more light than does a ⅛″ rod.

Rigid Spiders and Cells. The thick-vane types of spiders are easier to build, and they maintain alignment better. The vanes are typically ⅛″ thick, may be made of wood, metal, or plastic, and are attached rigidly to the central hub that holds the diagonal cell. To align the optics, the legs are moved inside the tube until the diagonal is properly positioned and angled. This is tricky because changing the mirror position also changes its angle. The advantage is that once adjusted, the rigid spider and cell will hold alignment indefinitely, so the legs may be glued to the interior wall of the tube.

Adjustable Spiders and Cells. Thin-vane spider legs are made of thin, flexible metal and are held under tension across the tube. Thin spider legs spread the diffracted light over a wider area and so provide better performance. For guided astrophotography, solidity is the most important requirement. Most commerical spiders are of the thin-vane type and are sold with matching diagonal mirror cells.

Commercial spiders and cells give you the quickest and easiest way to mount the diagonal mirror if you don't want to make the small, tricky parts for an adjust-

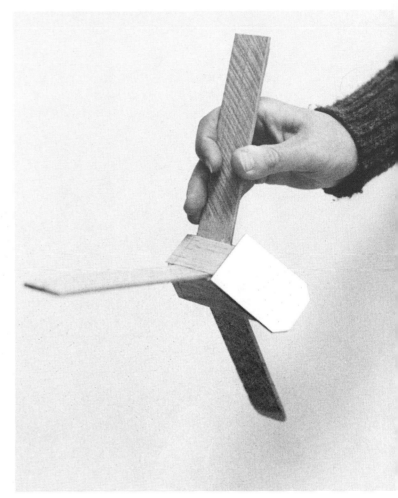

Rigid spider mounts are easy to build and very stable, but tricky to align.

able spider and cell. Check to see that the spider will fit, or can be adjusted to fit, the tube's diameter, and, especially for larger diagonals, that the cell will fit.

FOCUSERS

A focuser is a device for moving the eyepiece back and forth in search of the focal plane. It is desirable to bring the focal plane as close as possible to the telescope tube to minimize the size of the secondary mirror needed (the equations for diagonal mirror size show this; see Chapter 2). If at all possible, make or buy a focuser or holder less than 3″ tall, but plan on using it focused 1″ outward from its lowest position. The focal plane will then lie approximately 4″ from the tube.

Standard American eyepieces are mounted in 1¼″-diameter barrels intended to extend roughly 1″

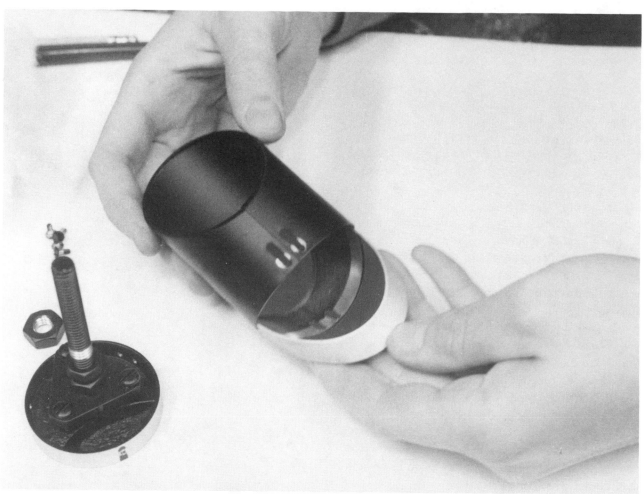

Installing a diagonal mirror in a commercially made adjustable holder; cotton batting presses the diagonal against the front of the holder.

From the biggest to the very simplest, take your pick of eyepiece holders.

into the eyepiece holder. The barrels are most often chrome-plated and have a smooth, hard surface that does not scratch easily. The holder should grip the barrel tightly enough that the eyepiece cannot fall out when the focuser is pointed down, but lightly enough that you can remove one eyepiece and insert another without disturbing the telescope's position.

Eyepiece Holders. Inexpensive holders grip the eyepiece with two small tabs bent inward in a thin-walled metal focuser tube. If the tube diameter is right and the tabs are bent correctly, this simplest of all holders does a remarkably good job. Imported holders use a thin-walled tube with a collar around it; a tiny thumbscrew threaded through the collar clamps the eyepiece. This scheme works well providing that the threads on the thumbscrew and collar don't get stripped and the tiny thumbscrew is not lost.

Better holders employ a thick-walled focuser tube and a plastic thumbscrew. The soft, plastic tip of the thumbscrew cannot damage the barrel of the eyepiece, and if lost, it can easily be replaced since it's a standard component available at well-stocked hardware stores.

A do-it-yourselfer who doesn't have a lathe can buy thin-walled 1¼″ inside-diameter brass tubing from a hobby specialty firm and then saw a T-shaped slot and bend the tabs slightly inward. An inexpensive (but not terribly good) alternative can be found at plumbing stores: 1¼″ sink-drain tubing made of chrome-plated brass; the flared end fits most eyepieces. With a metal lathe, of course, you could turn a focuser tube from any available stock, then drill and tap it to accept a thumbscrew.

Another inexpensive holder is the 1¼″ sink-drain tube to a plastic pipe adaptor. This accepts an eyepiece barrel when it's loose but grips firmly when tightened.

Friction Focusers. This simple form of focuser is composed of a tube holding the eyepiece; it slides inside another tube or a hole through a focuser block. It can be made from virtually any kind of material—plastic, Bakelite, or metal. The traditional friction focuser is made from two telescoping thin-walled brass tubes, the outer one soldered or brazed to a plate that mounts on the telescope tube.

A good alternative to brass tubing is ⅛″ wall Bakelite tubing available from plastics suppliers. Plastic plumbing pipe sliding in a soft-wood focuser block also works

reasonably well (and is compatible with the 1¼″ sink-drain adaptor used as the eyepiece holder). Again, your ingenuity is important.

For the super-low-cost route, use a piece of sink-drain tube as the eyepiece holder. Make a wooden focuser block for the tubing by drilling a hole just under 1¼″ diameter through a 1¼″-thick block of white pine. Mount the block on the tube of the telescope, and the focuser is complete.

Friction focusers seem to work best if you twist the tube slightly as you push. Pushing often knocks the telescope off of whatever you were looking at; twisting and pushing usually don't. It takes a while to acquire the knack.

Helical Focusers. These are the low-cost favorite among commercially made focusers. The focusing tube is not smooth but has a spiral thread machined on the outside that engages a spring-loaded ball bearing or other detent in the focuser block. These devices would seem to offer very fine focusing and rapid focusing by popping the threads past the ball, but in practice, they are frustrating to use. Neither fine nor coarse motion works well.

Rack-and-Pinion Focusers. This classic type of focuser is supplied on most commercial Newtonians. The device is basically a friction focuser with a few additions. A toothed track (the rack) is attached to the focus tube, and a small pinion gear fixed on a shaft with knobs is mounted in the focus block. By turning the knobs, the focus tube is forced back and forth. Well-made rack-and-pinion focusers perform satisfactorily; however, if the focus tube is slightly loose in the block, or if the pinion gear is loose in the rack, the tube rocks from side to side or jams because the tube is pushed from one side only. These problems are helped by increasing the tension on the pinion gear and lubricating the sliding tubes.

Crayford Focusers. An exceptionally smooth-focusing mechanism, this type works by guiding the focuser tube between rollers. It avoids the wobble and jamming problems of the rack-and-pinion focuser by allowing one, and only one, degree of freedom to the focus tube. Four small-diameter ball bearings are mounted in pairs on a post; the tube is pressed against them by a spring-loaded rod. As you turn the rod, the focus tube moves. Crayford focusers are extremely stable and sensitive to the touch and allow quick

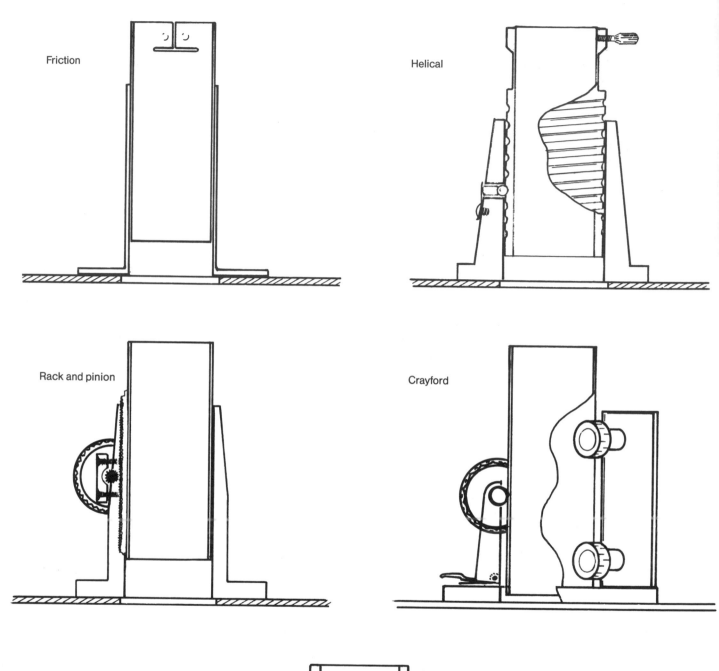

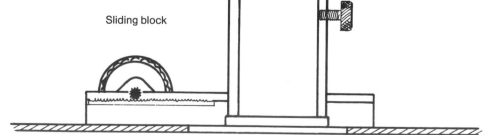

Five styles of home-built eyepiece holders.

changes from one eyepiece to another if each is mounted in a separate focus tube.

Sliding-Block Focusers.   These separate the eyepiece-holding function from the focusing function. The eyepiece holder can be very low, allowing for a smaller diagonal mirror. The entire eyepiece assembly slides lengthwise along the tube. Because the block can be longer and better supported than short telescoping tubes, these focusers tend to be smooth and steady. Their major disadvantage is that the diagonal mirror must travel with the block, so it must be mounted on a single stalk or two-stalk spider attached to the block.

Commercial Focusers.   Price is not a reliable indicator of commercial-focuser performance. A metal-and-plastic friction focuser costing under $10 may well outperform the handsome commercial rack-and-pinion focuser costing $60. Another rack-and-pinion unit may outperform both. The best way to choose a focuser is to inspect and test it before buying.

FINDERS

Locating objects with a telescope is difficult at any magnification above 30x, and frustrating or fruitless above 100x. A finder takes the place of setting circles and helps you locate hard-to-find objects as it gives you a nice preview of the sky surrounding the objects you're looking for.

Peepsights.   These allow you to point the telescope, but they have no optical power. The foresight is mounted at or near the top of the tube. It is a ring thick enough to be seen against a dark sky but not so thick that it blocks much, and it is distinctive enough that there is no ambiguity about where the object should appear relative to it. At or near the bottom of the tube is the hindsight ring, which the observer looks through.

The internal diameter of the foresight ring should be roughly 1/60 of the distance between the foresight and the hindsight; the hindsight ring is ¼″ in diameter. Hardware for this type of peepsight is readily

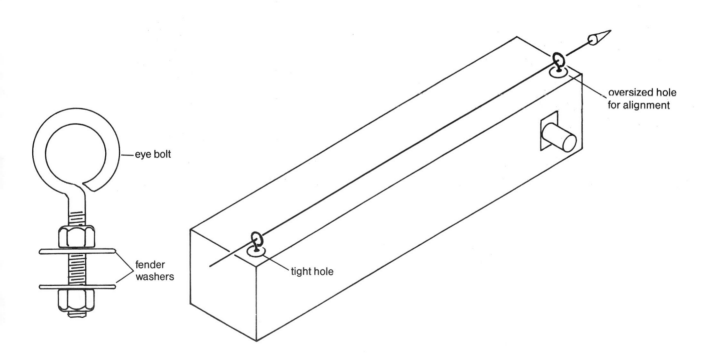

A peepsight is adequate for finding bright stars and planets at low magnification.

available—a large eyebolt is the foresight, and a screw eye the hindsight.

Mount the foresight firmly near the upper end of the tube; then cut and trim a small block of wood to hold the hindsight near the bottom end. After collimating the telescope, point it at a bright star, then move the hindsight until the star is centered in the foresight. Screw the hindsight down in that spot. Alternatively, both the foresight and hindsight can be eyebolts. The foresight is in a tight-fitting hole, but the hindsight has an oversize hole. Fender washers allow you to move the hindsight until the peepsight is aligned.

As stated before, a peepsight provides no magnification or light-gathering power, but if you're observing mostly bright objects, a peepsight may be adequate to your needs. One minor problem is that kneeling on the ground to look through the hindsight hurts the knees. In any case, you can start with a peepsight and use it for a few weeks until you get around to making or buying a finder telescope.

Finder Telescopes.   A finder having ¼ to ⅓ the aperture of the main telescope and a field of view covering 9 to 16 times the sky area of a low-powered eyepiece is a valuable addition to any telescope. Finders are amazingly easy to construct *if* you can get the materials. Buy an achromatic lens 50mm to 60mm in diameter with a focal length between 150mm and 300mm (surplus binocular objectives are the best source of these) and an eyepiece with a focal length of 25mm to 30mm. After a few years of amateur

astronomy, you'll probably have a spare eyepiece or two to use for a finder, or you can buy a low-priced one for the purpose.

The lens and eyepiece are mounted in a tube. Thirty minutes spent in looking at plastic plumbing components may prove invaluable here: A 1¼″ adaptor can hold the eyepiece; a cap or adaptor can hold the lens; and they can be joined by a length of plastic pipe. Glue the lens in place with silicone sealer, cut the pipe to length, put it all together, and then finish your new finder to match the rest of the telescope.

On reflectors especially, "straight-through" finders force you to turn your head at an uncomfortable angle to reach the eyepiece. Adding a diagonal mirror or prism to the optical path brings the light out the side of the tube. Right-angle finders flip the already-inverted field, making star-hopping more difficult, but the ease of use compensates for this problem. You may want to make several finders having different magnifications and fields for your telescope.

Finder Mounts.   Ring mounts can be made from plywood, plastic, or metal. They hold the finder and also space it away from the tube so the observer can get in line with the eyepiece. Four inches is usually enough clearance. One problem to watch for: People have an annoying tendency to loosen the bolts absent-mindedly. Your only protection against this is to use hex-head or socket-head bolts, which cannot be turned by hand.

Commercial Finders.   There are a few excellent finders and a lot of cheap imports on the market.

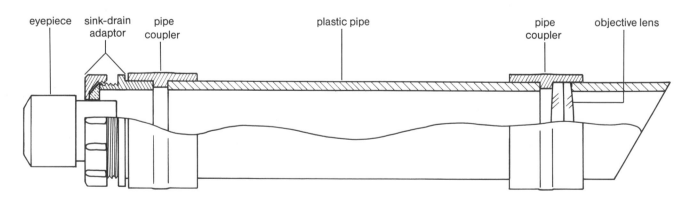

Cross section of a simple finder made from standard plastic plumbing components.

The best finders are made from components that are designed for finders, have standard 1¼″ eyepieces, and cost quite a bit. Import finders are made from standard, small import refractor components, employ 24.5mm-diameter eyepieces (usually Huygens or Ramsden), and cost only a little less than well-made ones. If at all possible, examine the product before buying it. If it provides a bright, wide field of view and seems reasonably well-made mechanically, and if you don't feel like building the finder, then go ahead and get it. One key diagnostic point: Unscrew the eyepiece holder from the tube and examine the threads. If they are shallow and ragged, it's one of the imports.

EYEPIECES

Choosing good eyepieces is a matter of considerable importance. Much as you might like to skimp on them—and eyepieces are expensive—this is one area where skimping does not pay. If you are working within a small budget, buy a few high-quality eyepieces for low magnification plus one or more inexpensive, short-focus eyepieces for high magnification.

The function of the eyepiece is to transfer to your eye the image formed by the telescope objective without messing up that image. Unfortunately, eyepieces aren't perfect. They may contribute spherical and chromatic aberration, astigmatism, coma, field curvature, and distortion of their own to the final image. The first four of these aberrations depend quite strongly on the telescope's focal ratio; the lower it is, the worse the performance of the eyepiece.

In order to purchase eyepieces intelligently, you'd need to know how much of each aberration a given eyepiece would produce with your telescope and then how to weigh that against its cost. This information, sadly enough, is not available from the sellers. The best you can do is to accept their word for the optical type and trust their reputation. Most eyepieces sold in the United States are jobbed out to foreign manufacturers. Reputable importers design, specify, and spot-check the wares they're selling; others order generic designs (which may or may not be well-designed and well-made) and sell them at a substantial markup. There is no better way than to try

Mount your finder on the telescope tube with ring mounts cut from plywood.

it before you buy it, or to buy only from firms that sell with a "money-back-if-not-satisfied" guarantee.

Of course the acceptable aberrations vary with the type of observing you will do. In planetary and lunar viewing, for example, axial image sharpness is important and field size less important. Observing star clusters, nebulae, and galaxies with an eyepiece that has a wide, sharp field adds greatly to the pleasure and excitement of observing, so you might willingly accept less-than-perfect central image sharpness for good definition over a large area.

However, there is no excuse for *poor* central image sharpness in any eyepiece. No matter what the observing task, eyepieces must be corrected for spherical and chromatic aberration in the center of the field. If your eyes are good, the telescope objective good, and the seeing conditions good, but you still can't seem to get a sharp focus on an object in the

Eyepieces come in all sizes and shapes but in only three standard barrel diameters: 24.5mm, 1¼″ (the most common), and 2″.

center of the field, the culprit is probably one of these two aberrations.

From roughly 15° off axis to the edge of the field, the image is usually dominated by astigmatism, field curvature, lateral color, or a combination of these. Since the amount and character of the off-axis star blur depends strongly on the focal ratio of the telescope, an eyepiece that performs reasonably well on an f/8 Newtonian may be a disaster with an f/5 Newtonian. While it may still deliver crisp images in the center, astigmatism can make stars at the edge of the field appear cross-shaped, comet-shaped, or even shaped like flying birds!

Strong field curvature is easy to recognize. When you focus the center, the edges look fuzzy; when the edges are sharp, the center is fuzzy. You can tolerate weak field curvature because as you look around the field, your eye automatically refocuses. Be that as it may, you'll get the impression that *something* is wrong because stars will look fuzzy until you stare directly at them.

Whether distortion will bother you or not is up to you. Distortion is a displacement of the image from its correct place—a rectangular grid imaged as a "pincushion" (the usual case with eyepieces), or sometimes as a "barrel." Distortion may be one trade-off the designer makes for freedom from coma and field curvature. In astronomical viewing, distortion is most noticeable when you move the telescope. The stars whizzing through the edge of the field appear to move along curved paths—something you can probably tolerate.

Another important feature of any eyepiece is the distance at which you must hold your eye behind it, or eye relief. This is especially important if you wear glasses. Too little distance means that you must remove your glasses in order to position your eye close enough to the eyepiece to see the whole field of view. The shorter the focal length of the eyepiece, the shorter the eye relief, so for high-powered viewing, you will almost certainly need to remove your glasses, no matter how much you need them. This causes no problem as far as observing is concerned unless you have extremely bad astigmatism because you can simply refocus the telescope for your eye. However, it drastically increases the probability that you'll misplace your glasses or find them by sitting on them in the dark.

Too much eye relief causes difficulty for people who don't wear glasses, especially with long-focus eyepieces at low magnification. If you're one of these people, you must then hold your eye some distance behind the eyepiece with nothing to guide you but the edges of the field vanishing from sight. Rubber eye cups help this. They not only keep out stray light, but give you tactile feedback in positioning your head.

Barrel Diameter. Eyepieces are available in three standard sizes: 24.5mm (0.965″), 1¼″, and 2″ barrel diameters. The 24.5mm barrel is typical of inexpensive import telescopes and generally indicates a poor-quality eyepiece. The 1¼″ is the standard American size, suitable for eyepieces with focal lengths of less than approximately 30mm. At this focal length, an eyepiece with a 60° apparent field of view has a field lens of about 30mm in diameter, nearly the largest feasible inside diameter of the barrel. Longer-focus eyepieces mounted in 1¼″ barrels necessarily have relatively restricted fields of view. A 2″-diameter barrel accommodates the long focal-length eyepieces needed for low-powered viewing with f/6 and slower telescopes.

Parfocal Eyepieces. Parfocal eyepiece sets slide into the eyepiece holder to a common focus. This means that you can change magnification without much refocusing. There is no difference in the optical design of a parfocalized eyepiece; the difference lies entirely in the barrel and in how far the eyepiece slides into the focuser. Unfortunately, not all parfocalized eyepiece sets are parfocalized to the same point, so one set of parfocalized eyepieces may not match another.

EYEPIECE TYPES

The pioneer telescope-builders of the 1920s made do with a miscellany of old microscope eyepieces and simple eyepieces they made themselves. Eyepieces were not available in focal lengths longer than 25mm; they had narrow fields of view, were nonachromatic, and uncoated. The coupling of low-quality eyepieces with the relatively long-focus mirrors and refractors of that era meant that observers in the not-too-

Two useful accessories: the right-angle, or "star," diagonal; and an "eyepiece adapter" for holding a 1¼″ eyepiece in a 2″ focuser.

distant past were stuck with too much magnification, tiny, cramped fields of view, and dull, fuzzy views of celestial objects.

Modern eyepieces are much better. Many are computer-designed for optimum performance with fast optical systems and are developed to give a wide field of view, high transmission, and little internal reflection. Furthermore, they're available in a wide range of focal lengths that will provide optimum magnification for the type of observing you're doing.

Huygens eyepieces were the standard fifty years ago; today they have virtually disappeared. The design has strong field curvature, usually yields only a 25° to 35° field of view, and introduces spherical aberration. While they are suitable with slow optical systems (i.e., refractors), they give terrible images with fast Newtonian reflectors. Huygens eyepieces were once popular because they were easy to make and inexpensive, and because better designs were not available. Cheap import telescopes often come equipped with Huygens eyepieces.

Ramsden eyepieces are a vast improvement over the Huygens and nearly as cheap and easy to make. Their spherical aberration is about 20 percent that of the Huygens', but some residual lateral chromatic aberration remains. The field of good definition is typically 35° to 40°. They perform moderately well with an f/8 or slower Newtonian but are poorly suited for f/4 and f/5 Newtonian RFTs. A Ramsden eyepiece consists of two elements: identical plano-convex lenses oriented convex to convex and separated by about three-quarters of their focal length. The effective focal length of the eyepiece is equal to the focal length of the individual lenses. If you feel like handcrafting every part of a telescope, Ramsdens are one of the few eyepiece types you can make for a reasonable amount of effort.

Kellner eyepieces are simply achromatized Ramsdens—the eye lens is a doublet rather than a single-element lens. Lateral color is greatly reduced relative to the Ramsden, but astigmatism and field curvature are still large. The field may be as much as 45°, although with that field, the edges will be fuzzy. Kellners are frequently supplied with bargain-priced telescopes and low-cost binoculars, yet they are often good enough to work well with amateur telescopes, particularly the "slow" ones.

The **RKE** is Edmund Scientific Company's computer-optimized variation of the Kellner eyepiece. In the RKE, the field lens is a doublet and the eye lens is a singlet. The field is 45°. Its performance exceeds that of the classic Kellner and approaches or exceeds that of Orthoscopic eyepieces. While they are not as good as well-made Plössls, they cost about half as much.

Orthoscopics were the "eyepiece of choice" in the 60s and 70s, their name virtually synonymous with optical perfection. Although they give a well-corrected 45° field with little lateral color, they suffer from astigmatism and a strongly curved field. Optically, the eyepiece consists of a single-element eye lens and a triplet field lens. For their price, Orthoscopics provide excellent value, especially for high magnifications.

Plössl eyepieces are a further development of the Kellner, consisting of two doublets (rather than a doublet and a singlet). They enjoy an excellent reputation among observers, giving essentially perfect axial images and considerably less astigmatism then the Kellners. Typical fields are 50°. Most observers agree that a well-made Plössl gives better performance than a well-made Orthoscopic, and although their distortion is greater, this usually does not matter much for astronomical use.

Erfles can't be beat for wide-field observing at a moderate price. They are designed to give a field of 60° to 65° across with reasonably good definition at the edge. There is considerable variation in the optical design and in the quality of manufacture. It is best to test the brand name, and preferably the individual eyepiece, before buying.

There are many other eyepiece designs, some merely variants on the standard ones. The **König**, for example, is a simplified version of the Erfle. Others incorporate as many as seven elements or incorporate strongly curved lenses that are hard to manufacture. They are expensive, of course, but perform superbly. Especially notable are the **Nagler** ultra-wide-angle eyepiece, which gives an 82° field of view with good correction (but costs roughly three times that of comparable Plössls), and Tele-Vue's **Wide-Field** eyepieces, designed to outperform Erfles but costing twice as much. Although it is not necessary to buy the most expensive eyepieces, especially when you're just starting out in a new hobby, be willing to spend enough to get good ones.

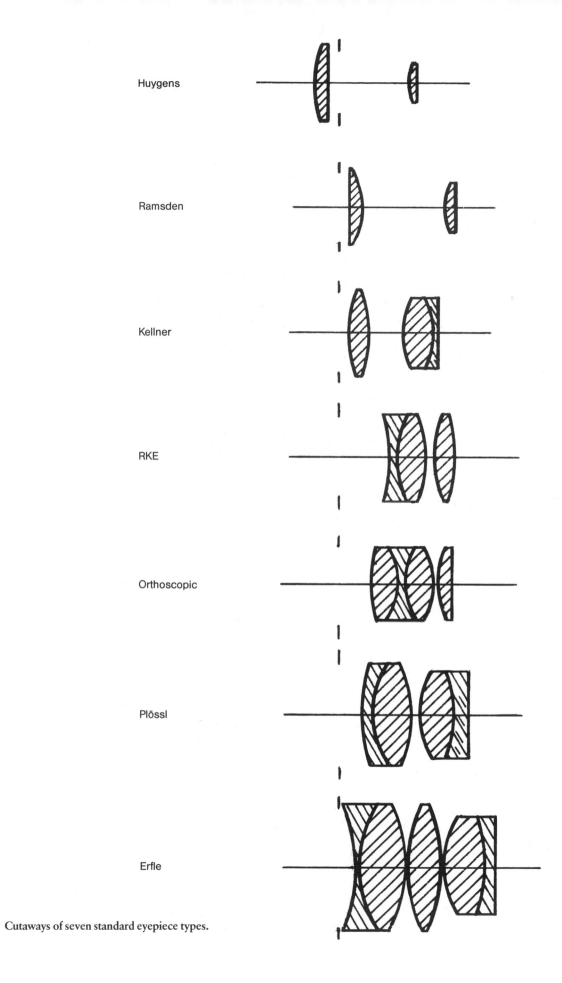

Huygens

Ramsden

Kellner

RKE

Orthoscopic

Plössl

Erfle

Cutaways of seven standard eyepiece types.

Before amateur telescope-making began in the 1920s, only the rich could afford a telescope. Russell W. Porter opened the door for thousands of would-be astronomers with his article "The Poor Man's Telescope," which explained the delicate art of shaping glass for home-built telescopes.

Today you have the choice of whether to make or buy. Buying the mirror means that you can start observing sooner. "After all," you might argue, "how many biologists grind the lenses for their microscopes? A telescope is a *tool* for me, not an end in itself."

Making the mirror, of course, means that your telescope will be much more *your* telescope—a source of pride and proof of a skill acquired. You'll understand how it works and know the true meaning of precision. It means that you will not be limited to the sizes and focal ratios sold, so you can make unusual telescopes for specialized observing programs.

Home-Brewed Optics
10

HOW BIG?

Since the 1920s, the 6″ f/8 mirror has been the standard beginner's mirror for the very good reason that he is virtually assured of success providing he sticks with the job and doesn't give up. The glass is easy to handle, and the stages of grinding and polishing speed by. Polishing, figuring, and parabolizing, despite all the literature devoted to them, are not very critical operations in this size and focal ratio. The difference between a spherical and paraboloidal mirror is just large enough that you can't leave a 6″ f/8 primary spherical.

An 8″ mirror is not much larger, and it is certainly no trouble for an adult to handle. Since the difficulty of parabolizing increases by the square of the aperture and the inverse square of the focal ratio, an 8″ f/6 mirror is roughly three times more difficult to figure than a 6″ f/8, but that's still nothing to be afraid of. If you know someone who has made mirrors before and is willing to give you his or her personal tutelage, an 8″ mirror would be a good starting point.

Larger sizes call for more developed skills. When you perform a given figuring stroke on the surface of a 6″ or 8″ mirror, the whole surface changes. Larger mirrors act differently. When you fix a problem here, a new problem pops up over there. Unless you have

Russell W. Porter (right) and Albert G. "Unk" Ingalls founded the amateur telescope movement in America. Photograph by Robert E. Cox.

the experience from smaller mirrors (plus the conviction based on previous success that you will ultimately lick the mirror's problems), you can spend a lot of time learning basics you should have learned earlier.

There's an old saying that goes, "If you want to make a 12″ mirror, make a 6″ first. It'll take you longer to make a 12″ straight off than to make a 6″ and then make a 12″." It's true; you need the experience you gain from a small mirror to make a large mirror.

MATERIALS

You'll need the following materials:

Mirror Blank.   This is a disk Corning #7740 Pyrex glass or another low-expansion, borosilicate glass. Depending on the supplier, the blank may be a casting (for which molten glass was poured into a mold)

with characteristically sloping sides, ridged front, and lumpy back; it may have been cut from sheet glass and have straight, ground sides and a nearly flat front and back; or it may have been pressed into a mold while soft and have characteristically nubbly surfaces.

Most blanks are approximately one-sixth as thick as they are in diameter to ensure that they will not bend under the forces of grinding and polishing, or under their own weight in a telescope. But because Pyrex has become expensive, several firms sell "thin-mirror" blanks for large mirrors. Thin blanks are more difficult to grind and figure; they should be avoided by beginners.

The Tool.   This is another piece of glass the same size as the mirror. It is often simply another mirror blank or, depending on the supplier, a disk of plate glass or ceramic material. The tool need not be as thick as the mirror, but it should not be less than half as thick as the blank. A second blank used as a tool has the benefit of potentially serving as a mirror blank after it has been used as a tool.

Abrasives.   You will need a sequence of abrasives, starting with coarse and finishing with fine. The normal sequence is #80, #120, and #180 Carborundum, followed by #220, #320, #400, #600, and #800 Alundum, and, finally, #305 emery, although this sequence varies in kits. Grains of these sharp, hard substances roll between the blank and the tool, grinding by chipping away the glass. Carborundum is the harder of the three and is better for coarse grinding. Aluminum oxide is less expensive and so is normally included as a fine abrasive in mirror-making kits.

The number assigned to an abrasive indicates its "mesh size"; that is, the number of wires per inch in the screen the material sifts through. The coarsest abrasive used in mirror-making, #80 Carborundum, is a little finer than table salt but has a distinctly gritty feel. Fine abrasives, such as #305 emery (an abrasive listed by stock number rather than mesh size), feel finer than flour.

Keep the various grades of abrasives—ranging from coarse #80 "Carbo" to flour-fine #305 finishing powder—separated from each other in sealed containers.

Abrasives are cheap in 25-pound drums (usually the smallest commercial quantity sold), but a 6″ mirror needs only one pound of #80, a half-pound of #120, four ounces of #180, and two to three ounces of each of the rest. For an 8″ mirror, buy twice those quantities. If you join an astronomy club or visit the mirror-making shop at a large planetarium, you can purchase small quantities at a good price. If not, small containers are available from mail-order suppliers at prices that reflect the labor involved in repackaging and shipping it.

Optical Pitch.   Pitch is an extremely viscous liquid made from tree resins or petroleum; it flows slowly when warm, acts like a solid at room temperature, and shatters like glass when hit hard. You use it to make a "lap" for polishing. The pitch is melted and poured onto the tool in a thick layer.

Pitch flows to conform exactly to the surface of the blank in several minutes, but over the time interval of a few strokes, it acts rigid. Pitch allows the polishing lap to stay in contact with the blank even as you alter the shape of its surface with figuring strokes.

There are many varieties of pitch, differing mostly in hardness. Not all of them are suitable for making a fine optical surface. Buy pitch from a firm that sells it for amateur mirror-making; or buy it through a club that

has made a bulk purchase of the right stuff. One pound is enough for a 6″ mirror and two pounds for an 8″, but it doesn't hurt to order an extra pound or two in case you must make a new lap halfway through polishing.

Polishing Agent. This used to mean red optical rouge—an excellent, but incredibly messy, polishing material. Nowadays most amateurs prefer cerium oxide (or brand-name polishing agents based on it) because it polishes faster and is less messy than rouge. Rouge, however, gives a better optical surface. As is the case with abrasives, small quantities are relatively expensive. You'll need four to six ounces for a 6″ mirror, and eight ounces for an 8″.

The items listed above can be obtained together in a mirror-grinding kit. (Because most of the kits on the market cost less than the component parts purchased separately from the same supplier would cost, they are extremely convenient for an isolated beginner.)

Grinding Stand. You will need a sturdy, hip-level work surface on which to support the blank and tool. Ideally, you should be able to walk completely around the stand, but the corner of a solid kitchen table

will do in a pinch. A simple box 16″x 16″ by 32″ to 36″ high, depending on your height, made of ¾″ plywood and filled with sandbags or concrete blocks so it won't rock or tip, works well and is easy to construct.

Grinding Board. Attached firmly to the stand is a grinding board, a square of ¾″ plywood about 16″ on a side. Screw to it three blocks of clear pine ½″ thick, 2″ long, and 1″ wide. Two of the blocks should be placed with their inner edges tangent to a circle the diameter of the tool and blank, and the third block placed ½″ radially outward from the circle. Also make a soft pine wedge 3″ long, ½″ thick, and ¼″ at the narrow end and ¾″ at the wide end. The blocks and the wedge hold the mirror during grinding.

Paint the entire grinding board with several layers of sealer/primer and a coat of light-colored floor enamel; then let it dry thoroughly.

Carborundum Stone. You must bevel the edge of the blank to prevent chipping. A Carborundum knife-sharpening stone roughly 6″ long, 2″ wide, and ½″ thick, made with sintered #120 or #180 grit, is the perfect tool for this.

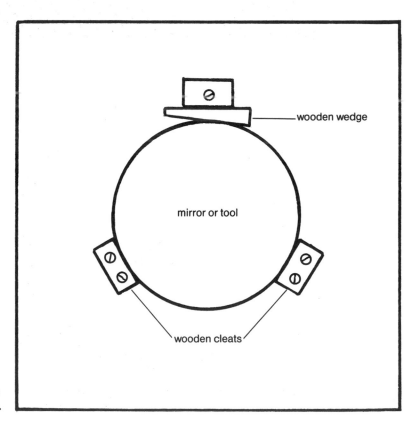

Wooden cleats on the grinding board hold the mirror during grinding and polishing.

Squeeze Bottles.   One-pint plastic squeeze bottles, the type used for dispensing mustard and ketchup, are good for adding water to grit or polishing agent to the lap. Buy two, and label one "gritty" and the other "polish" with an indelible marker.

Small Pot and Hot Plate.   For melting the pitch, you'll need a one-quart saucepan, either enameled steel or aluminum, and a hot plate with covered heating elements. An old pot and an electric stove will do, but melting pitch and casting the lap is a very messy job. Warning: *Do not heat pitch over any type of flame—flammable vapors boil off and could catch fire.*

Single-Edged Razor Blades.   These blades are the best tools for trimming the facets in pitch laps. They have a reinforcing strip along the back edge. Do *not* use ordinary two-edged blades—they are too dangerous! Razors are sharp enough to cut brittle pitch but cheap enough to throw away as soon as one blade gets the tiniest bit dull. Buy a box of 100 at a hardware or hobby shop.

Fiberglass Window Screen.   You need only a square foot of this for pressing a small pattern into the pitch lap before a polishing or figuring session. Do not substitute metal screen.

Foucault Tester.   You must construct a Foucault (named after Leon Foucault and pronounced FOO-coh) tester and a mirror stand to hold the mirror while testing it. Both are described in Chapter 11.

Logbook.   For recording every work session, write down what you did with what and for how long. This will be especially valuable when figuring.

There are other necessary items that you probably already have on hand: a plastic drop cloth to prevent contamination, an apron so you don't get glass and Carborundum "mud" on your clothes, a small scrub brush, masking tape, duct tape, and a roll of plastic food wrap. In addition, it's nice to have a metal straight-edge, a feeler-gauge or a large set of small drills for measuring the depth of the curve while rough-grinding the mirror, but if you don't have them, don't go out and buy them.

THE WORK AREA

It's important to keep the work area scrupulously clean so that no coarse abrasives, chips of glass or metal, or anything else that could scratch a smoothed optical surface can get between the blank and the tool while working. Optical work also requires stable temperatures.

Kitchens or laundry rooms are often good work areas. They are usually clean, or at least have cleanable surfaces; running water is easily available; and temperature stability is good. Your working there, though, may interfere with the activities of other family members.

Garages make terrible work areas since they are filled with windblown dirt, metallic debris, and concrete dust unless the door is always kept shut. There is usually no water either, and the temperature varies wildly.

Basements are good work areas, though often they're not clean. Their particular problem is that all kinds of grit drop from the ceiling. The easiest and most practical solution to this is to staple a plastic drop cloth to the rafters and tape a sheet of 4-mil plastic on the floor. Place the grinding stand centrally on the plastic and put a plastic bucket of water to one side. A small table nearby can hold the abrasive, water, and other items you need while working.

In the polishing stages, cover the grinding stand with a sheet of plastic when you're not using it. This protects the worktable from people who might leave gritty objects (such as flowerpots) on it. Be aware that the plastic cover may itself become a source of contamination if a grain of #80 abrasive hidden on the grinding stand sticks to it, then is transferred to the grinding board and the blank. Even if that were to happen, good work habits, which we'll discuss later, should prevent damage to the optical surface.

PREPARING THE BLANK

Select the side of the blank that is to be the surface of the mirror. The sides may be nearly identical, in which case the decision hardly matters; otherwise, choose the flatter, smoother face. Also inspect for bubbles bigger than $\frac{1}{32}''$ within $\frac{1}{4}''$ of the surface (if the surface is not clear, wet it); avoid using a side with large bubbles near the surface.

Throughout the grinding and polishing, the disk must remain chamfered with a $\frac{1}{8}''$-wide 45° bevel around the circumference of both sides of the blank

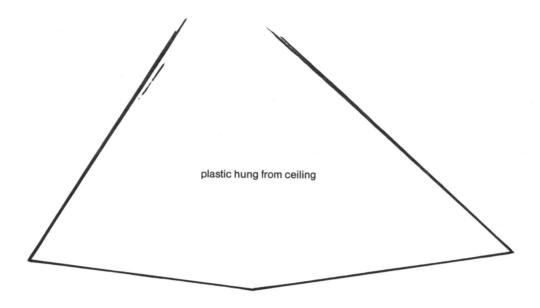

plastic hung from ceiling

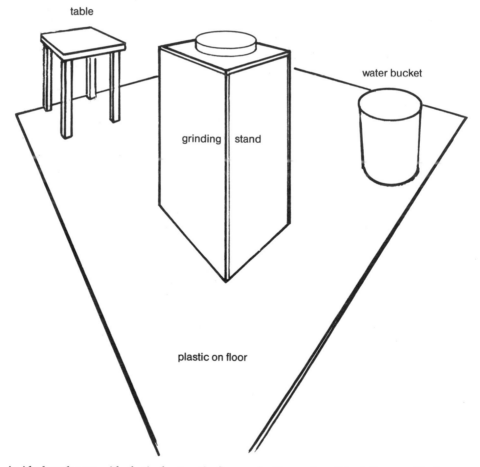

table

water bucket

grinding | stand

plastic on floor

An ideal work area, with plastic sheet on the floor and ceiling and a solid grinding stand in the center.

and tool. This prevents edge-chipping. A large chip means starting over again with a new blank.

Use a Carborundum stone to form the chamfer. Dip the blank in water, then hold it so that it hangs over the edge of the grinding stand. Dip the stone in water, hold it at a 45° angle against the blank, then lightly stroke away from the blank several times across the edge. This will grind off some of the glass. Rotating the blank, work around it once lightly. After the first pass or two, increase the pressure, but if you see tiny chips forming, reduce the pressure. Work around the blank again and again until the chamfer is ⅛″ wide. Keep both the blank and the stone wet. Water acts as a lubricant and coolant and removes the fine particles of glass that are ground off.

Cast blanks sometimes have a slightly raised rim on the "front" side; you should remove this rim before starting. Grind it off against a piece of ¼″ plate glass about twice the diameter of the blank. Use long, straight strokes with #80 abrasive and water, working until the rim is approximately level with the rest of the surface. Rough grinding will take off a lot more glass, but selectively from the center of the disk. You would prefer not to have to worry about whether or not the rim or the low zone just inside it will grind out in time. If the back of the blank has sharp ripples or bumps on it, grind it flat; #80 "Carbo" knocks superficial bumps off very quickly. Keep an eye on the chamfer during all grinding; restore it whenever necessary.

ROUGH GRINDING

In rough grinding, there are three goals: 1) to remove material from the center of the blank and the edge of the tool so that the blank becomes concave and the tool becomes convex; 2) to control the depth of the curve; and 3) to bring the tool and blank into contact over their whole surfaces so that both are sections of matching spheres.

Fill the plastic bucket and the "gritty" squeeze bottle with water, put the #80 Carborundum in an accessible spot, and place the tool on the grinding board. Now, before starting, a note about cleanliness: #80 is a coarse abrasive, and you'll use a lot of it. If a single grain of #80 gets between the blank and the tool

during fine grinding or polishing, it will scratch the mirror and cost you hours of work. Avoid getting grit in your hair or clothes. Don't throw it around. Be neat.

Wet the tool with a small amount of water and sprinkle about ⅓ of a tablespoon of #80 on it. Then place the blank on the tool and grip it firmly with both hands. Applying moderate pressure, push the blank forward an inch or two, then pull it back an inch or two past center. You'll hear a loud, gritty sound as a thin layer of abrasive rolls between the blank and the tool, breaking off tiny flakes of glass, and breaking down itself. Rotate the blank a fraction of a turn, take a small step around the stand in the opposite direction, then push the blank forward again.

Continue to grind and "walk around the barrel" (as it's called in old telescope-making books) for a minute or so until the abrasive's action and the accompanying grinding sound weaken. You will find a grayish mixture (called "mud") of broken abrasive grains and glass chips between the glasses. The interval taken for the abrasive to break down and the cutting to cease is called a "wet."

At first, the motions of grinding may seem jerky

Chamfer the rim of the mirror and tool with a Carborundum stone.

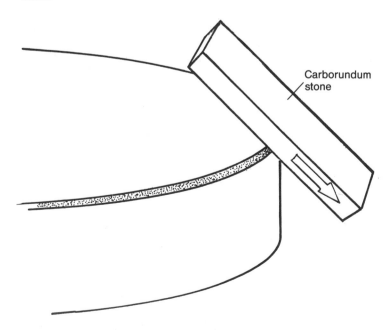

Carborundum stone

Sprinkle a moderate amount of #80 "Carbo" on the wet tool surface.

and irregular. Do not stop or worry about it—just keep on grinding. With practice, you'll develop an easy rhythm that you could keep up for hours at a time, a motion that is smooth and regular. "Walking around the barrel" ensures that the grinding is distributed evenly around the tool, and the little turn of the blank in your hands ensures that, over a period of time, the grinding action will occur around the blank evenly. Do not resist the natural, random variations from stroke to stroke; they'll make sure that the surface will be smooth and blended.

Swirl the blank in the bucket to remove the milky-gray mud of abrasive and glass chips, rinse the tool, then do another wet. Experiment with the amount of

water and abrasive. If there is too much water, the abrasive runs off before it grinds anything; too little water and the blank won't slide over the tool. If there is too much abrasive, it is pushed off the edge of the tool before doing any work; too little and the small amount of abrasive present is broken down very quickly. Once you find the best amount of water and abrasive to use, wets will be noisy and efficient.

Now you can begin removing glass. A long center-over-center stroke will produce a concave blank and convex tool, but slowly. To take off glass quickly, concentrate your grinding on the edge of the tool and the center of the blank with a chordal stroke. Apply water and abrasive to the tool, then place the blank on the tool with its center about three-quarters of the radial distance to the edge of the tool. Take a short stroke tangent to the edge of the tool, forward and back. Rotate the blank slightly, take a step around the stand, and then take another stroke forward and back.

Don't try to work too fast—40 forward and back strokes per minute is plenty. This allows you to maintain a good pace for a long working session, and later, working too fast will distort the mirror's figure. Apply plenty of weight—be serious about removing glass—but keep the weight evenly distributed. Under no circumstances should the side of the blank overhanging the tool tip off; it could chip the tool or the blank. The entire overlapping section should grind, although the emphasis will, of course, be on the outer edge of the tool and the center of the blank. Continue grinding with this stroke until the abrasive is exhausted, rinse the blank and tool, add new abrasive, then continue. Under these conditions, the abrasive breaks down quickly and the wets are short.

After 30 minutes or so, a cavity will begin to develop in the blank. The relation between the focal length of the mirror and the depth of the curve is:

$$depth = r^2/4F$$

where r is the radius of the blank and F is the focal length. For a 6″ f/8 mirror, r = 3″ and F = 48″, so the depth of curve should be 0.047″.

To measure the depth, place a straightedge across the mirror, then determine the size of a drill or feeler-gauge that will slip under it. Start with small drills

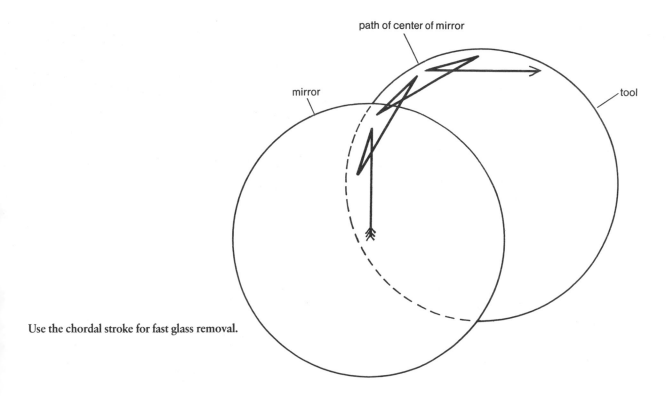

Use the chordal stroke for fast glass removal.

Slow strokes and heavy pressure will shape the mirror and tool quickly and efficiently.

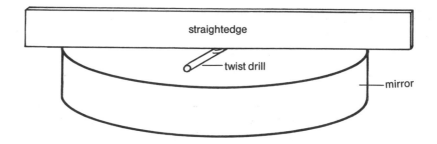

Gauging the depth of the mirror's curve with a twist drill of known diameter.

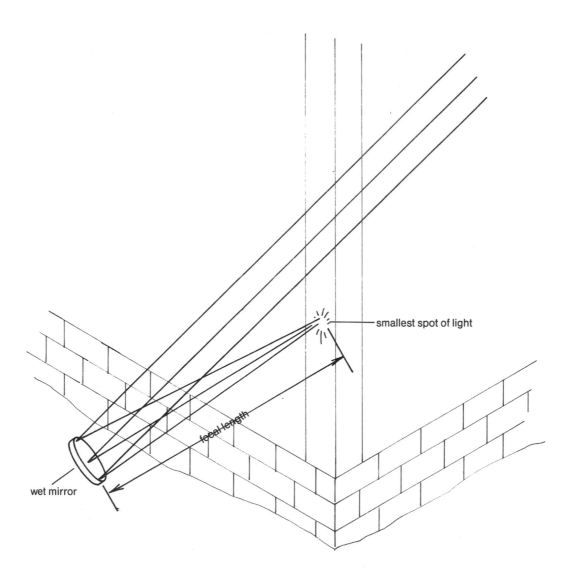

Estimating the focal length of a wet mirror by focusing sunlight.

and work up. If the diameter of the largest drill that slips under the straightedge is D, then:

$$F = r^2/(4D)$$

If you don't have a reliable straightedge or a big set of drills, use an optical method instead. Take the mirror outside into sunshine, prop it up facing the sun, then splash it with a little water with dish detergent in it (the detergent makes the water flow smoothly and evaporate more slowly). Use a sheet of cardboard to catch the crude solar image formed by the water. You can measure the blurry "best focus" within 6″.

If the day is cloudy, stand the mirror on edge in a darkened room, wet it, and, holding a flashlight beside your head, shine it at the mirror. Move around until you catch the beam of light reflected from the wet surface. Move your head back and forth. If the light moves in the *same* direction as you move, you are *closer* than twice the focal length, so step backward and try again. If the light moves in the *opposite* direction as you move, you are *more* than twice the focal length from the mirror, so step closer and try again. Keep the mirror wet. When the light brightens and dims but does not move from side to side when you move, you and the flashlight are twice the focal length from it.

When the focal length is roughly 60″, begin "spherizing" the surfaces. Continue grinding with the mirror blank on top, but switch from the fast-working chordal stroke to a milder "W" stroke. Begin with a long center-over-center stroke, but make the next strokes farther to the left and shorter until the center of the blank is three-fourths of the way to the edge of the tool; then take each farther to the right, cross the center, and zigzag over until the blank's center is three-fourths of the way to the right side. The path of the blank's center in this stroke is a big, zigzaggy "W." All of this, of course, is done while you're still rotating the blank and walking around the barrel!

As the mirror crosses the center, you may feel a pull because the out-of-spherical tool and blank do not fit each other. If unbroken abrasive accumulates in the center, stop grinding and redistribute it with your fingers. As the fit between the surfaces improves, the tool and the blank will begin to grind each other everywhere.

Use the "W" stroke until the focal length is 48″ as best as you can tell. While the "W" stroke will not have produced spherical surfaces, the surfaces should no longer be grossly distorted as they were from the chordal stroke. Switch to the "normal" stroke: one-third of the diameter, center-over-center—the one you will use throughout fine grinding and polishing. With a 6″ mirror, the total stroke length will be 2″ and the edge of the blank will overhang about 1″ at maximum excursion. Alternate wets with the blank on top, then the tool on top. This makes the surfaces fit without changing the focal length. Each blank-on-top wet shortens the focal length slightly, but the tool-on-top wet that follows will lengthen it again.

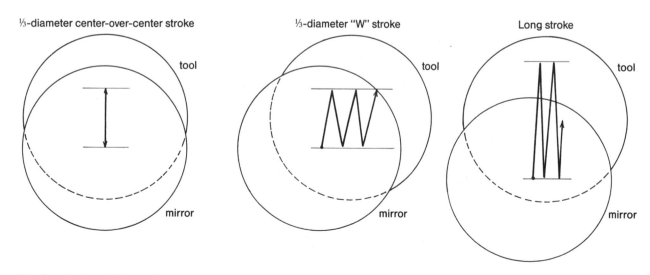

The three basic grinding strokes.

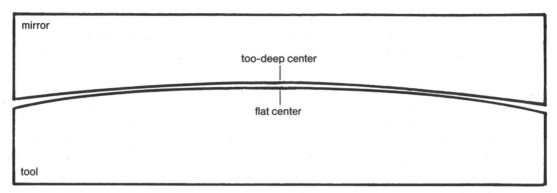

mirror

too-deep center

flat center

tool

After rough grinding with the chordal stroke, the surfaces of the mirror and tool will be distorted.

After each of the four or five alternating sets of wets, measure the focal length as carefully as you can. If it is too short, grind two wets with the tool on top for each wet with the tool on the bottom; if it is too long, grind two wets blank-on-top for each wet done blank-on-bottom. Do not be overly fussy. Errors of 4″ or so can be corrected during fine grinding, and you probably can't measure the focal length much more accurately at this stage.

While grinding, watch the behavior of the bubbles and the abrasive slurry between the mirrors. If bubbles tend to accumulate in the center and the abrasive breaks down slowly there, the tool and the blank are not in good contact, so continue grinding with the normal stroke.

When both focal length and contact are right, the rough grinding is finished. The whole job, start to finish, should have taken about six hours. Clean up all traces of #80. Unscrew the grinding board, take it outside, hose it down, and scrub it thoroughly. Scrub both the mirror and the tool, the rinse bucket, and anything else that may be contaminated with coarse grit.

FINE GRINDING

There are three goals in fine grinding: 1) to remove pits left by coarser abrasives so you end with a surface ready to polish; 2) to bring the blank and tool into increasingly better contact with each other; and 3) to complete the process of fine grinding with the desired radius of curvature.

In fine grinding, you use a one-third diameter, center-over-center stroke, grinding alternate wets blank-on-top and tool-on-top. As the abrasives become finer and finer, the duration of a wet increases and the number of wets required to remove pits from the previous abrasive decreases.

Cleanliness is extremely important. Between every change of grit in fine grinding, unscrew the grinding board, take it outside, hose it down, and scrub it again. Also scrub the stand, re-cover the small table you used for grit and water with fresh newspapers or plastic, and scrub the rinse bucket thoroughly inside and out. Wear a shirt with short sleeves when working. Be careful, also, that the outsides of your abrasive containers do not become contaminated or contaminate the water squeeze bottle. Store each can of abrasive in a separate zip-lock plastic bag.

Finally, before starting work, scrub the grinding board, mirror, tool, bucket, and squeeze bottle once again (better safe than sorry). Rinse your hands in the water bucket each time you touch potentially contaminated objects; grit will wash off and sink to the bottom of the bucket.

Begin with #120 Carborundum. Its wets last for two to three minutes. After the first two wets, dry the blank and inspect its surface with a magnifying glass (note: Inspections will be a potential source of contamination throughout grinding, but you can't get along without them; rinse your hands and the mirror before resuming grinding after an inspection). The surface should already look finer, although large #80 pits will remain.

The #120 should have ground the entire surface. If it did not, you may have misjudged the fit of the

Inspect the surface for pits left over from the previous grade of abrasive.

blank and the tool. If there is any doubt in your mind, keep a close watch on the distribution of air bubbles and abrasive between the blank and the tool. If the fit is good, bubbles and abrasive will not collect in the center.

Keep the grinding stroke smooth and rhythmic, without sudden jerks or catches when you reverse its direction, at 40 to 60 strokes per minute. Don't strive for machinelike perfection. Variations in length, rotation, and timing will average out to produce a much smoother surface than the most perfect machine could produce.

Grind another ten wets, then reinspect the blank. No obvious #80 pits should remain, although you will probably find a few other abnormally large-sized pits. Mark their locations with a grease pencil on the back of the blank. Also check the focal length. If it is too long, continue, but do two mirror-on-top wets for each tool-on-top wet; if it is too short, do two

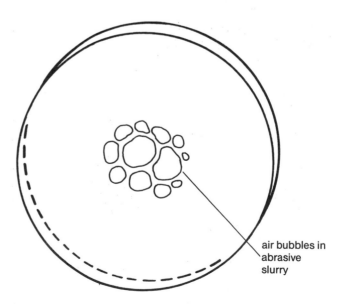

air bubbles in abrasive slurry

Bubbles collecting in the center are bad news: They mean the mirror and tool are not spherical.

SCHEDULE FOR FINE GRINDING

Grit Size	Wets*	Time (mins.)
120	40	120
220	32	100
320	20	100
400	16	80
600	12	80
800	10	60
305	8	60

(*Note: Rely on the absence of larger pits.)

tool-on-top wets for each mirror-on-top wet. If it is close to 48″, alternate whichever one is on top.

After ten more wets, inspect for pits again. The original large pits should now be gone, and a few new ones you had not seen before will have appeared. Mark their locations, then do another ten wets. If once again the biggest pits are gone and new ones have appeared, and if the focal length is close to 48″, you are finished with #120.

Each succeeding abrasive grade is treated in much the same way, although you will find that the length of a wet increases and the number of wets necessary decreases. Reduce the pressure you use in grinding too; after #320, apply just the weight of your hands, and for the final finishing abrasive, only the weight of the mirror. With finer abrasives, a smaller amount of abrasive is necessary because the distance (and therefore the volume) between the blank and tool is smaller. The focal length should be within an inch of 48″ (which is about as accurate as you can measure the focal length of a wet fine-ground mirror) when you finish #320.

Fine grinding is the best time to remove pits without grief. If there are residual pits, they'll stand out clearly when the mirror is polished—and you can't polish out large pits. Even though they don't do much harm to the image quality, they look terrible. If you have any doubts, do another half-hour of wets just to be on the safe side.

Scratches often occur as you begin a new wet; you

sprinkle on the abrasive "flour," lower the other disk on top, give it the first stroke, and "screak"—it's too late. You can catch most stray grains if you work as follows: Rinse both disks, put the appropriate one on the bottom, rinse your hands, squeegee excess water off with the side of your hand, sprinkle on a small volume of abrasive, spread it around with your fingers (you can feel any abnormal particles), then gently lower the other disk on, listening for any crunching. If you hear something, lift the disk straight off, with no lateral motion. Rinse off the contaminated grit, sprinkle on new abrasive, and try again. This procedure takes only a few extra seconds.

If the blank does get scratched, try to guess which abrasive grade will remove the damage. (You should be familiar with the pit-size made by each abrasive in the sequence, so use the next finer grade than your estimate of its size to remove the pit.) Return to that grade, grind until the scratch is gone, then repeat all the intermediate steps.

Don't let the abrasive slurry dry out; be especially on guard against this if you live in a dry climate. You may find that in order to simply keep the work wet enough you must add a few drops of water or start a new wet even before the old abrasive is exhausted. If the disks show any sign of sticking or grabbing, separate them immediately. Despite these precautions, sometimes a too-dry pair will lock together. If prolonged soaking in warm water or pounding the palm of your hand against them doesn't loosen them, have someone else hold the pair on edge, place a short length of 2″x4″ against the protruding disk to act as a cushion, then whack it with a leather mallet. There is, obviously, some risk of breaking or chipping one or both disks.

You may notice that as you grind the surface finer, the glass surface begins to reflect light—enough so that you can see the ruddy image of a bare light-bulb filament when you hold the blank so it catches the light at a low angle. Inspect the surface for uniformity: Rough areas reflect less than smooth areas do. Once again, pay special attention to the edge and center.

When you have completed the finest abrasive, the surface will be clear enough to read large type through when dry, and virtually transparent when wet. The glass will have a peculiar sticky-smooth feel. If the surfaces are in good contact, they should be within a few wavelengths of a perfect sphere. Furthermore, if

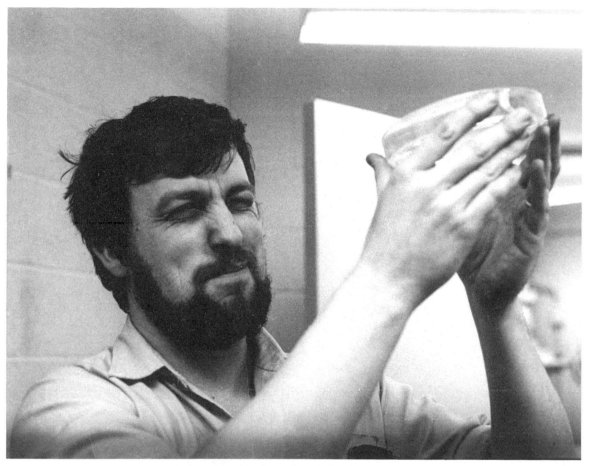

A properly fine-ground surface reflects light when viewed at a low angle.

you have been careful and attentive, even the most scrupulous inspection will turn up no residual pits.

When the last abrasive is finished, wash everything in the work area thoroughly. Scrub the grinding board, clean the water bucket, and store everything related to abrasives—containers, Carborundum stone, the "grit" water bottle—in individual plastic bags.

MAKING A PITCH LAP

Pitch is exasperating stuff. It smells awful when you melt it, flows when you want it to solidify, shatters when you try to trim it, and sticks to everything it touches. Nonetheless, the properties of pitch allow you to shape the mirror precisely. A lap made of pitch adjusts itself to the shape of the glass over a span of a few minutes while acting rigid during each stroke.

For precision optical work, the lap is faceted, or divided, into small squares that flow and adjust quickly. To form a lap, cover the tool ⅛″ to ¼″ thick with melted pitch, let the pitch partially solidify, then press the mirror against it to mold the pitch to the glass. Then, while the pitch is still soft, cut channels into its surface with a sharp razor blade. Finally, remold the pitch to conform to the mirror's curve.

To make a lap, cover a table or counter top with newspapers and set an *electric* hot plate nearby for melting pitch. Access to running water is necessary. The kitchen is an excellent work place, but remember that this is a messy job. Again, melting pitch gives off flammable vapors—*never heat pitch over an open flame.*

What else do you need? A small pot, a spoon, a container of water at room temperature, a smooth-edged coin (a nickel), a small quantity of pure turpentine, your trusty plastic bucket (scrubbed clean again, both inside and out) for water, paper towels, a roll of 1"-wide masking tape, single-edged razor blades, a plastic ruler about 1" wide. Also prepare separating fluid: Fill a small jar (such as a baby-food jar) one-quarter full with polishing agent, then add water, shake until creamy, then add a tablespoon of dish-washing detergent and stir it in well.

Place one pound of pitch in a small pot over *low* heat. While it's heating, fill the plastic bucket with warm water and place the tool and mirror in it to warm them up. After a few minutes, take them out, add more hot water, and replace the glass. Keep doing this until the water is uncomfortably hot to the hands.

Stir the melting pitch often so it remains homogeneous. If it starts to bubble, it's too hot—remove the pan from the heat and turn down the heat. When the last pieces finally melt, dribble a sample into the container of room-temperature water. It will solidify into a small blob. Let it stay in the water for 30

Pouring a pitch lap.

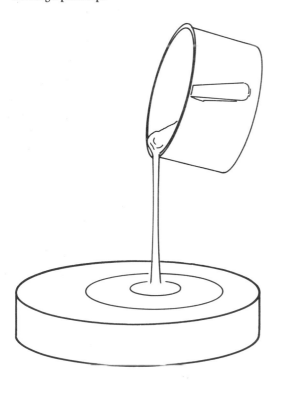

seconds, then retrieve it and place it on a hard surface. Hold the coin at a 45° angle against the bead and press it with the weight of your relaxed arm and hand.

If the pitch has the proper hardness, it should take you between 30 and 45 seconds to press a dent ¼" long into the bead. Soft pitch will be almost gummy in feel, and the dent will form in 10 to 15 seconds. Pitch that is too hard will show a ¼" dent in 60 or more seconds but will shatter glasslike if you press hard.

Old-time opticians would place a bead of pitch in their mouth until it warmed up to body temperature, then gently chew it. If it shattered between their teeth, it was considered too hard; the right pitch behaved like hard taffy, yielding but not getting soft like chewing gum. This was obviously an acquired art.

There is a broad mid-range of acceptable pitch. If the pitch seems okay, leave it be. If it is too soft, continue heating it to evaporate some of the solvents; retest it every ten minutes or so. If it is too hard, remove the pot from the heater and temper it with a quarter-teaspoon of pure wood turpentine (not paint thinner). Then stir well and test it again.

Pitch that has been "cooked," or tempered, is never as responsive as pure pitch of the proper hardness. Once again, if it seems to be okay, leave it be.

Remove the pitch from the heat; let it begin to cool. Remove the tool from the water bucket and towel it dry. Pour a little turpentine on a clean paper towel and scrub it vigorously on the face of the tool. This cleans off any grease that might prevent the pitch from sticking to the tool. Wrap a "dam" of masking tape ¼" high around its rim, then place the tool faceup in the center of your work area.

Remove the mirror from the bucket. Place it faceup beside the tool. Pour on about a teaspoon of separating fluid and smear it around with your fingers. This soapy mix prevents the pitch from sticking to the mirror.

By now, the pitch in the pot will be thickening. When it reaches the right thickness, slowly pour it onto the center of the tool. (Later, after doing many laps, you will know from experience exactly how viscous pitch should be for the best results; for your first try, pour when it's about as thick as molasses.) The pitch will spread slowly toward the edges of the tool. Keep pouring slowly so that it reaches the dam all around at about the same time. Despite being warm, the tool is still cooler than the pitch, so the pitch will flow more slowly at the edge of the tool. Since you want a convex lap, slowly pour more pitch in the center. When you have a layer roughly ³⁄₁₆″ thick at the edges and ¼″ thick in the center, stop.

Check to see that the separating mixture on the mirror is still wet and slippery by smearing it around with your fingers; add more if necessary. Touch the pitch gently with a wet finger; light pressure should just dent the surface. Pull off the masking-tape dam around the tool, then pick up the mirror and place it facedown on the pitch. To prevent sticking, move it slowly with small, circular strokes and no pressure. If your judgment and timing were perfect, the mirror will contact the pitch at the center and spread it smoothly toward the edges.

If the pitch was too warm and soft, it may chill and adhere weakly to the mirror, sliding around on the warmer pitch below. *Do not* lift the mirror off—keep it on top and hope that when the deeper pitch cools, you will be able to free the mirror. You may end up with wrinkles all over the lap, but don't worry

about them.

If the pitch has hardened, bear down on the mirror. It may take considerable time and elbow grease to work the lap into contact. If the lap gets too cold, warm the tool and mirror in hot water, add fresh separating mix, and keep working.

If the pitch was not sufficiently convex and contact began at the edge, you must get rid of the air trapped in the center. Place the center of the mirror over the edge of the pitch lap and work around the edge until you can slide the mirror across the center in contact.

When the mirror and tool are fully in contact, slide the mirror off one side of the lap, then trim away the pitch that has overflowed the edge of the tool. Allow the lap to cool for several more minutes, smear more separating mix on the mirror, and reestablish contact.

At this stage, you may be a wreck. However, while the pitch is still relatively soft and unlikely to shatter, cut facets into the lap with a sharp razor. The pattern of facets must be offset from the center by a different amount in each direction. Locate the center, then lay the plastic ruler about ¼″ to one side of it. Run the razor along the edge to mark this line. Turn the ruler

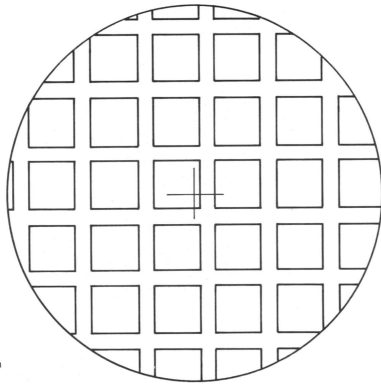

The pattern of facets should not be centered on the mirror.

A pitch lap ready for polishing; this lap has a hex pattern rather than the easier-to-cut square pattern recommended.

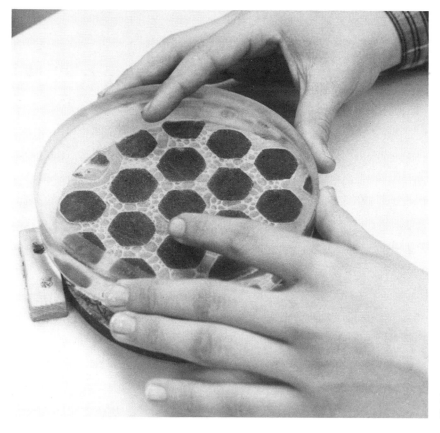

Light pressure and a slow stroke are best for polishing.

90°, offset about ⅝″ from center, and mark this line. Mark a third line ¼″ from the second, then score both lines several times with increasing pressure. The pitch between them should fracture.

Repeat this procedure for the remaining channels. When the blade gets dull, use a new one—sharpness is essential. Use the razor first to score the surface, then to chip out a shallow channel and, finally, to shave channels down to the glass surface of the tool. Keep the razor angled away from the facets so they don't chip, and try not to distort the surface of the lap. Do the best you can; don't let a few chips bother you unduly.

When you have finished faceting, trim away the pitch that overflowed the side of the tool. The face of the lap should be slightly smaller than the mirror blank. Trim it just as you trimmed the facets.

Finally, rinse the pitch chips and shavings off the lap. Smear the lap and the mirror with separating mix again, place the mirror on top, and establish contact. Then set them aside and clean up the mess you've made of the kitchen!

POLISHING OUT

Polishing proceeds much as fine grinding did above, but with the tool pitch-coated and the wets lasting from five to ten minutes. The basic polishing routine consists of a normal one-third-diameter, center-over-center stroke with alternate polishing spells done mirror-on-top and tool-on-top. By now, of course, rotating the work in your hands and walking around the barrel is second nature to you. Depending on the pitch, the pressure and peculiarities of your working habits, the fine-ground blank—now rapidly becoming a "mirror" —will polish out in four to six hours.

Make a polishing mix by combining 6 heaping table-spoons of polishing agent, 8 fluid ounces of water, and a drop of liquid dish-washing detergent in a clean glass jar. Shake it thoroughly, then let it sit for an hour to settle out any coarse particles. Pour off the top three-fourths of the mixture into the second squeeze bottle. Since what's left in the jar is probably not contaminated, make your next batch of polishing mixture on top of it.

Begin each session with a 30-minute "cold-press."

Add a fresh charge of polishing mixture, put the mirror on the tool under a moderate weight (i.e., around ten pounds; a gallon jug of water works well), and let it sit. The pitch will slowly flow into contact with the mirror.

A new lap takes a little breaking in. Start polishing slowly and gently, with the mirror on top, getting a feel for the lap. If the mirror suddenly "hangs" on an irregular facet or dry spot, stop, cold-press for a few minutes, then try again. If you haven't established good contact after several tries, warm the mirror and tool in a tub of lukewarm water to soften the pitch, recharge the lap with polishing mixture, "hot-press" for several minutes with fiberglass window screen between the mirror and the lap and a ten-pound weight on top, then begin polishing. The screen produces very small facets which press into contact quickly.

After a few minutes of work, the lap starts to "drag" —not with the sudden grab of poor contact, but with a steady, heavy pull indicating that the lap has flowed into intimate contact with the mirror. Notice that the facets look dark; the polishing-mix layer between the lap and the mirror is so thin that it's almost invisible. Increase the pressure on the mirror to overcome the drag; polishing has begun in earnest.

Maintain the steady, rhythmic, normal stroke as long as you can. It should not be necessary to stop except to add a small amount of fresh polishing mixture or water from time to time, and to alternate the mirror and tool positions every ten minutes or so. If you take time out, add a little polishing mixture and center the mirror on the lap.

The mirror will show signs of polish after five to ten minutes of work, but you might as well keep on polishing for an hour or so once you've gotten contact. At the end of that time, rinse the mirror clean and inspect it. Under average conditions, it will have become fairly transparent, and there will probably be a light "grayness" over its surface. If you have done your job well up to now, there will be no large pits and the semi-polish will extend to the very edge.

You must now put in several more hours of work. Polishing is complete only when the surface is invisible, or nearly so, even under strong illumination. Test it by shining a bright flashlight on it in a darkened room, or with the image of the sun formed by a magnifying glass. When you can't see either one anywhere on the surface, polishing is complete.

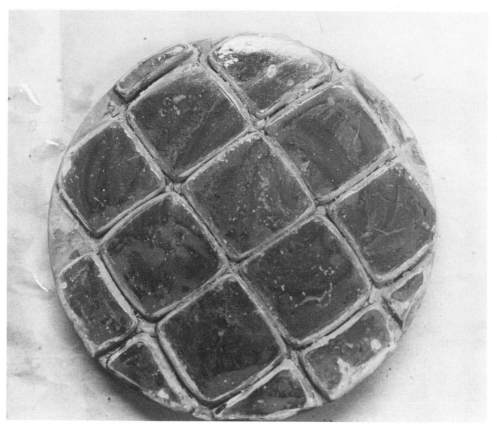

The channels have nearly closed; it's time for retrimming this lap!

Between daily polishing sessions, spread extra polishing mix on the mirror and leave the lap resting on it without any additional weight. To prevent the pair from drying out, place them in a plastic bag. Maintain the lap—if the pitch channels close to half their original width, heat the lap in hot water and trim the facets with a single-edged razor blade. Rinse away all shavings, recharge the tool with polisher, and leave the lap to cold-press until your next polishing session. If you will be unable to work for several days, store the mirror and tool separately in clean plastic bags, without any weight on the lap. Recharge the lap and cold-press it before starting work again.

FIGURING

Converting the "raw" polished surface into a precision paraboloid is known as "figuring." Testing is an integral part of figuring. You work, then test, then work, then test—again and again until the surface is right. Grinding and polishing are manual, but testing is primarily conceptual. Each test supplies information on the state of the mirror's surface, its current "figure." The next step, the real mind-bender for a novice mirror-maker, is converting test information into a plan for altering the surface.

Test results are expressed in terms of a reference surface, so mirror-makers speak of "holes" and "hills." Since figuring involves wearing away the high spots, to revise the "test figure" into a "working figure," you must express the surface error entirely in terms of "hills." You select a figuring stroke and carry out that operation for some period of time, then retest—and the process starts all over again.

In figuring, the stroke alters the rate of polishing in certain areas. The amount of material removed from a given part of the mirror depends on three variables: 1) time, 2) pressure, and 3) speed. Time is the most easily controlled quantity—to remove more, work longer. It is simple to control pressure by pressing harder with one hand than with the other, in effect, using one side of the lap rather than the whole tool as the polisher. It is best to hold the speed of your strokes constant. Work slowly—20 to 30 strokes per minute—so that the mirror maintains good contact with the lap at all times and a large number of strokes blends equally.

Just as you kept a log book for grinding, keep one for figuring. After a few test and figuring cycles, you will have forgotten just what has worked and what has not and whether or not it worked as you thought it would.

FIXING FIGURE ERRORS

You've just placed the mirror on the test stand, set up the Foucault tester (described in Chapter 11), and examined and interpreted Foucault shadows for the first time. What do you see? What do you do about it?

Spherical Surfaces gray uniformly. If you've never tested a mirror before, you may overinterpret insignificant errors on a surface that is nearly perfect. Don't overreact; if the mirror is good, you've done things right and can proceed to parabolizing. But before doing this, mount the mirror in the telescope and perform a star test on it (see Chapter 11) so you know what spherical aberration looks like.

Rings and Zones result from a lap that is symmetric or from a stroke that repeats too regularly. In either case, some zones have received more action than they should have. Inspect the lap. Is the middle facet centered, or do the channels in one or both dimensions fall symmetrically about a diameter? If the answer is "yes," make a new off-centered lap.

If the effect is not obvious at first glance, polish for an hour with the one-third-diameter, center-over-center stroke, but add a little randomness to the stroke length, and introduce moderate side-to-side excursions into the stroke. If these do not work, make a new lap.

If instead of many irregular zones, you find one raised ring, remove it by placing the raised zone tangent to the edge of the tool and polishing for five minutes with a short chordal stroke. Depressed rings are much harder to cure. The best method is to work for another hour, tool-on-top, with a more random stroke than before, then test to see if you've lessened or removed it.

Surface Ripple, or "dog biscuit," comes from polishing with a too-rapid stroke, or a lap that is not pressed

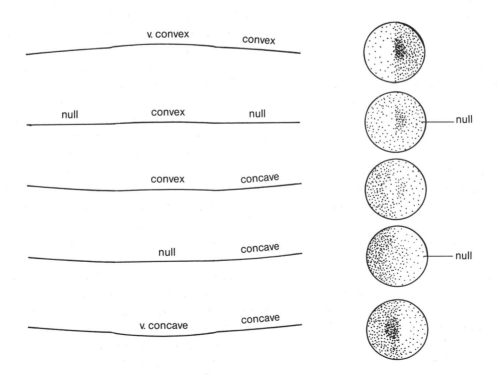

"Holes" and "hills" are relative terms in mirror testing, depending on which zone "mills."

Apparent Profile

Corrective Stroke

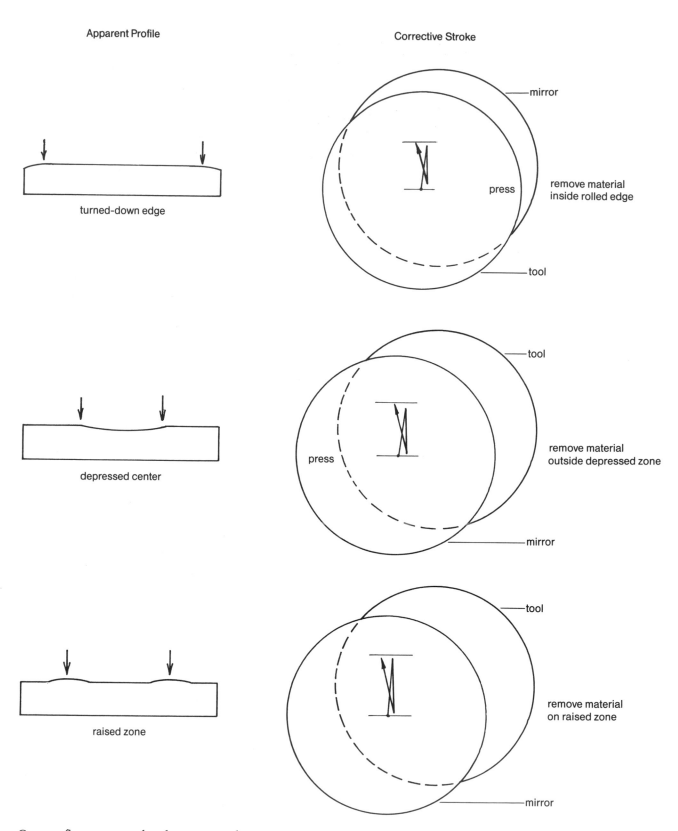

turned-down edge

mirror

press

remove material
inside rolled edge

tool

depressed center

tool

press

remove material
outside depressed zone

mirror

raised zone

tool

remove material
on raised zone

mirror

Common figure errors, and strokes to correct them.

into good contact, or as a result of prolonged figuring. It consists of dozens of small lumps and holes distributed randomly on the surface and is roughly the same size as the facets on the polisher. Dog biscuit shows up in the Foucault test as small variations in tone, and in the star test as scattered light, i.e., abnormally bright diffraction rings surrounding the Airy disk.

Surface ripple is difficult to avoid on short focus mirrors, especially once you've started figuring. The short work sessions and surface alterations with fast-acting strokes add up to a pattern of random ridges. In severe cases, dog biscuit results from a faulty lap or polishing technique. The cure is to make a new lap, then polish very slowly (15 to 20 strokes per minute) with a normal stroke while maintaining good contact.

Central Holes and Hills upset mirror-makers more than they should because the Foucault test is especially sensitive to small errors near the center of the mirror. The central 25 percent of the mirror's radius contains only 6 percent of its area; thus it should be considered relatively unimportant when compared to the outer 25 percent of the radius, with 44 percent of the mirror's area. Nonetheless, central defects are easy to fix. If the center is raised, displace the mirror so it overhangs the edge of the tool, and polish with a short "W" stroke. Use moderation: A little of this stroke goes a long way!

A central hole is equivalent to a raised zone outside the hole and tapering toward the rim of the mirror. With the mirror on top, displaced so the high zone is over the edge of the tool, work with a short stroke, blending the action outward with a "W" stroke. Again, do this work with moderation.

Turned-Down Edge (TDE) is the classic bugaboo of amateur telescope mirrors. TDE is bad because a large percentage of the mirror's area may be involved even if the turned region is quite narrow. It results from prolonged mirror-on-top polishing with a too-long stroke. Since you have worked with the mirror alternately above and below, it is unlikely that you will see TDE in either the Foucault or star test. To correct it, imagine the mirror surface as having a ring just inside the turned zone. With the tool on top, tangent to the "high" zone, apply extra pressure to the tool edge and work the edge around the mirror with a short "W" stroke.

If you see the classic doughnut shadows in the Foucault test, or pure, regular spherical aberration in the star test, run a series of zonal knife-edge readings on the mirror and plot the Foucault graph to determine how near a paraboloid it is; then skip to the next section.

PARABOLIZING

The tried-and-true method of producing a paraboloidal mirror is a two-step process: First attain a sphere; then "parabolize." This is an effective technique for beginning mirror-makers since a 6″ f/8 mirror departs only slightly from a sphere. Parabolizing a 6″ f/8 on a medium-hard lap with cerium oxide is simple. From a spherical surface, polish for five minutes, using the normal one-third-diameter, center-over-center stroke to get the lap warm and in contact. Then apply 72 three-quarter-diameter "W" strokes with a three-quarter-diameter side-to-side excursion. (If you are using optical rouge rather than a cerium-oxide-based polishing agent, do 200 strokes.) One stroke counts as a full forward-and-back cycle. It will take only two or three minutes to do this.

Wash the mirror and evaluate it either by making zonal readings and plotting a Foucault graph or with the star test, preferably both. If the figure is good according to the graph, shows smoothly blended Foucault shadows, is reasonably free of zones or dog biscuit, and displays no spherical aberration in the star test, the mirror is probably ready for aluminizing. But before you decide your work is finished, allow a few days for your jubilation to wear off, reread the chapter on testing, then test again with a freshly critical eye. If it still passes, it's really finished.

If the mirror's new figure is not a paraboloid, determine *in what way* it is not correct. It is for this, especially, that the combined powers of the star test and the Foucault test are most useful. The Foucault test gives quantitative guidance, and the star test tells how the mirror really performs on stars.

At this stage, you're *close*, so have the patience to work slowly. Be satisfied with one figuring step and a complete evaluation of the mirror per evening's work. Determine whether the figure is an ellipsoid (departing less from a sphere than a paraboloid) or a hyperboloid

The parabolizing stroke, and variations for retouching a defective paraboloid.

Normal Parabolizing Stroke

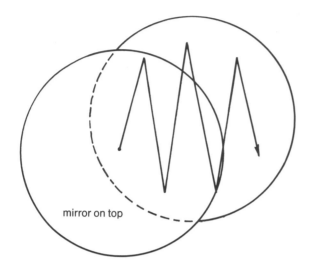

mirror on top

Deepening the Edge

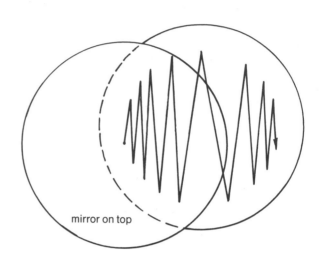

mirror on top

(departing more from a sphere than a paraboloid). The star test will show spherical aberration in both cases—make sure you determine whether it's over- or undercorrected.

If the mirror is ellipsoidal, simply continue parabolizing.

If the mirror is hyperboloidal, how much "over" is it? Do four five-minute sessions of the normal stroke, alternating the mirror position top and bottom, then retest the mirror. (The normal stroke is a very gentle method of "reducing" a hyperboloid.) If it has now become a sphere or an ellipsoid, apply the "W" parabolizing stroke but prorate the number of strokes for a full paraboloid. If the mirror is still hyperboloidal, continue the normal stroke until you reduce it to a sphere, ellipsoid, or paraboloid.

Chances are, however, that the figure errors will be trickier than simple over- or undercorrection. If the outer zones are flat (i.e., have too long a focus), use the parabolizing stroke modified with a shorter side-to-side excursion as the mirror crosses center; this is to concentrate the polishing action at the edge. If the inner zones are too shallow (i.e., have too long a radius), use the parabolizing stroke modified by taking shorter side-to-side strokes when the mirror overhangs the most; this is to put extra wear

in the center.

It's possible, but don't *expect* to get a good parab-oloid on your first try. By the third or fourth try, your previous experiences, if recorded in a log, should be quite useful in calibrating future parabolizing. If you consistently overcorrect, reduce the number of strokes. You may need to go back to a sphere (with an hour of normal strokes, mirror alternating on top and bottom) several times. Don't let these setbacks bother you—you're learning, after all, and probably closer than you think to being done.

There are several schools of thought on declaring the job completed. One school holds that you should stop as soon as the surface comes within ¼-wave (the "easy" Foucault graph criterion), and therefore is "good enough." This is typical of the quality you will find in commercial optics, and it's quite a reasonable criterion providing you understand that a ¼-wave mirror is acceptable but not excellent.

Another school of thought demands that you not stop short of perfection. What you lack to accomplish this, however, are test methods sufficiently delicate to detect tiny errors of figure. The Foucault test reads to roughly ¹⁄₁₀-wave for novices and half of that for skilled amateur opticians. Although the star test is quite sensitive, nights when the seeing is good enough

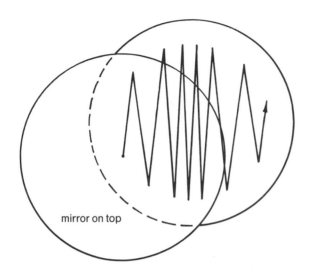

Deepening the Center

mirror on top

to show ¹⁄₂₀-wave errors are rare. The fact that the star test won't reveal the errors shows there is little to gain from striving to correct them.

Also, remember that the longer you mess around applying various figuring strokes, the lumpier the surface of your mirror becomes. Sooner or later you will begin to impair the optical performance you have been working so hard to achieve.

A demanding—but still practical—criterion for your first mirror is to stop when the zonal readings lie well within the tolerance envelope on the Foucault graph, and a visual examination of the Foucault shadows show them to be smooth and regular; or when you see a close match between the intra- and extra-focal images in the star test; or both. Make your second mirror perfect.

ALUMINIZING

After figuring, the mirror must be coated with a thin layer of reflective metal. Prior to the 1940s, chemically precipitated silver was used for most telescope mirrors. Although fresh silver is highly reflective (roughly 95 percent), within a few months, a ruddy-brown coating of silver-sulfide tarnish dulls its brilliance.

Overcoated aluminum coatings are preferred today.

Unfortunately, applying aluminum plus an overcoat is too complex for doing at home; instead, you must ship your mirror to a firm that aluminizes mirrors for amateur astronomers (see Appendix B). For a 6″ mirror, this will cost between $20 and $30. The mirror will be chemically cleaned, then placed in a vacuum chamber and the air pumped out. Some aluminizers run a high-voltage glow discharge to break up organic molecules still adhering to the glass, then pass a current through a tungsten wire covered with 99.999 percent pure aluminum. Aluminum evaporates from the hot wire, its atoms flying into the vacuum. When the atoms strike the cool glass, they stick. As soon as the layer is thick enough, the filament is shut off. Next, a "boat" containing silicon monoxide is heated; this compound also evaporates and deposits on the mirror. A silicon-monoxide "overcoat" is a clear, hard layer that protects the rather delicate aluminum coating from mechanical and chemical attack. The mirror is then removed from the chamber, wrapped in your original packing materials, and shipped back to you.

For shipping, wrap the face of the mirror with three or more layers of heavy plastic food wrap, enclose it in a strong, close-fitting box (preferably wood) tight enough that it cannot move, then repack that box in a larger cardboard box with plenty of crumpled newspapers. Be sure to include your return address inside both boxes as well as on their outsides. Most aluminizers will return your mirror to you within a month.

A freshly aluminized and overcoated mirror has a reflectivity of between 85 percent and 88 percent. Although aluminum is not as reflective as silver, the coating will last indefinitely if kept clean and dry, or five to ten years when exposed to the level of dirt and moisture found in a typical observatory (versus a few weeks or months with chemically applied silver).

Today, considerably higher-reflective coatings are also available: Overcoated "enhanced" silver, for example, is 98 percent reflective and reasonably durable. These coatings cost two to three times more. The additional cost may not be worth it for the single reflection from a single mirror (85 percent versus 98 percent), but for telescopes with three or more mirrors, the reflectivity of three aluminized mirrors would be $(0.88)^3$, or 68 percent, versus $(0.98)^3$, or 94 percent, for enhanced silver.

Optical Testing
11

How good is your telescope mirror? Can it do everything a telescope of its aperture should do? Can it really split those close double stars, show the Red Spot on Jupiter, Saturn's rings, and tiny lunar craters? Although it's guaranteed "excellent," how do you know it is? Are you, in fact, getting the quality and performance you paid for? How can you tell? In this chapter, we explore two methods of testing your telescopes's mirror—star testing and Foucault testing.

Star testing involves examining the image of a star produced by the mirror. This is done by putting a high-powered eyepiece in your telescope and examining carefully the inside-focus and outside-focus images of a moderately bright star. If they are exactly the same, the telescope has excellent optics. Star testing is something of a go/no-go test. Although it lets you easily recognize a problem, the star test allows only an eyeball estimate of the problem's magnitude. You can readily apply the star test to lots of telescopes—indeed, any telescope you look through.

The Foucault test is a geometric survey of the glass surface. It measures the precise location of light returning from different parts of the mirror and compares the measurements to geometric theory. You can then deduce from these measurements how to figure the mirror to correct the errors. This test is performed indoors in the optical shop; it yields both qualitative and quantitative results. You can estimate the "smoothness" of the figure by eye and readily detect zones and see abrupt figure errors. On the quantitative side, a graph of the zonal measurements shows a cross section of the figure.

STAR TESTING

The most practical test of any telescope is its performance on stars. What's best about testing on stars is that anyone can evaluate a functioning telescope without using any auxiliary equipment and, in a matter of minutes, determine whether its optics are excellent, good, so-so, or terrible.

What the star test reveals is limited, however, to certain yes/no questions: Does the telescope show collimation errors? Does it show any sign of astig-

The Airy disk and diffraction rings of a perfect telescope.

matism? Does it have spherical aberration? Is the mirror stressed in its mounting? Is the edge turned? Is the surface smooth? ("No" is of course the desired answer in each case.)

To star-test a mirror that's not yet finished, place it in its cell and tape it down snugly with masking tape, then install the cell in the telescope. Point it at a bright, uniform surface, then look down the focuser tube. Align the diagonal mirror so that the image of the primary is centered in it, then turn the mirror-alignment bolts until the diagonal mirror's silhouette is centered in the light reflected from the bare glass of the primary. This is not a precise method of alignment, but it's quick and will suffice.

If the telescope is one that's already finished, align it as above, then point it at any medium-bright star—Polaris is a good choice because it doesn't move—or, for testing an unaluminized mirror, point it at the brightest star available. Put in an eyepiece of 6mm to 12mm focal length, and center the star. Now carefully turn the focus knob back and forth about ⅛″. When it's in focus, the star should look like a tiny disk—the Airy disk—encircled by a few delicate diffraction rings. If you turn the focus knob slightly either way, however, it will appear larger, possibly displaying a system of concentric rings inside it, a dark spot in the center (the shadow of the diagonal mirror), and silhouettes of the spider legs. In a telescope with good optics, *a defocused star disk appears identical inside and outside focus.* In a telescope with even slightly defective optics, you will see deviations from this ideal.

The stellar disk may appear to do nothing other than expand (i.e., it might seem totally bland and featureless) on either side of focus, but look more carefully. Spend a few minutes: Focus back and forth, look, and let your eye accommodate to the image. You'll begin to see more. Roll through focus very slowly, then very quickly—snap, snap—to compare the inside- and outside-focus disks. If the extrafocal image seems to be shimmying, or if there are irregular blobs moving slowly across it, the air inside the telescope has not yet cooled down. If you see a rapid pattern moving in one direction across the mirror like flowing water, it's air turbulence, or what astronomers call "seeing."

To get some feel for how sensitive the star test is, hold your hand under the front of the telescope tube while looking at a bright, out-of-focus star. You'll see the silhouette of your hand and blobs of air warmed by it rising across the defocused star image.

INTERPRETING STAR-TEST IMAGES

Coma and Critical Collimation.   When you first test your telescope, chances are that the optics won't be accurately aligned but simply installed and eyeball-centered. The first two steps in star testing are letting the telescope cool down and then making a careful alignment. The principal aberration in misaligned Newtonian reflectors is coma—a "hairy" extension of a star image away from the optical axis. After a careful job of centering the mirrors, the effects of coma will be much more subtle.

Carefully run through focus looking specifically for any asymmetry in the brightness of the extrafocal disk; expand it until you see five or six diffraction rings. If the image is brighter, or if the extrafocal diffraction rings appear to be "bunched up" on one side and the asymmetry remains *on the same side* through focus, it's coma. (A defect in the mirror will switch sides on opposite sides of focus.) At best focus, coma appears as a protuberance or a flare of light sticking out of the star image and might pass unnoticed; the extrafocal image is sensitive to very slight misalignment.

The bright side of the comatic image always faces the optical axis. While examining the star, place your hand near the front of the telescope's tube, then move it around the tube until its silhouette appears on the bright side of the extrafocal image. Turn the adjustment bolts in the mirror cell to tilt the mirror toward the side where your hand was; this moves the image toward the optical axis. Repeat the test and align again. As you "tune out" the last of the coma, test for it closer to focus—two or three rings in the extrafocal disk. It shouldn't take more than four or five tries before achieving near-perfect collimation.

Astigmatism and Stress.   Next, check to see that the star image is round in both inside and outside focus. If the image is slightly oval on one side of focus, check the other side. If it's still oval but the long axis has rotated 90°, it has astigmatism from 1) your own eye, 2) the eyepiece, 3) the diagonal mirror, or 4) the primary mirror. Start by turning your head through the largest angle you can turn it; if the oval rotates with your head, the problem lies in your vision and should be corrected with glasses. If not, turn the eyepiece about its axis. If the oval rotates with it, the eyepiece is defective. The third possible source is the diagonal mirror. If it is not mounted too tightly (and thereby, warped), it's possible that this mirror is not truly flat, therefore defective. The easiest way to test a diagonal is to order another and substitute the new one for the one in question. If the image remains the same with the new diagonal, the problem is probably not in the diagonal.

The possibility remains that the primary itself is astigmatic. This is rare, for unless the mirror is made from unannealed plate glass, the random rotations in fine grinding and polishing virtually eliminate any chance of astigmatism. However, adjustable metal mirror cells encourage people to "get a good grip" on the mirror—such a good grip that they may warp the glass! Luckily, this is nearly impossible in silicone adhesive-mounted mirrors. Nonetheless, if there is persistent astigmatism and you cannot eliminate other causes, demount the mirror and remount it temporarily at a different angle. If the astigmatism rotates with the mirror, the mirror is astigmatic. You should carefully examine your grinding procedures if it's your own mirror; or, if you bought the mirror finished, return it as defective.

Checking the Edge.   Now concentrate on the outside edge of the extrafocal image. The rim normally looks

brighter than any of the diffraction rings, but it should be brighter by the same amount on both sides of focus. If it's brighter outside focus than inside focus, and if inside focus a haze or spray of light radiates from the disk, the mirror has a turned-down (long radius) edge.

If you see these effects but don't have any other telescopes to compare them with, place over the mirror a cardboard mask with an opening an inch smaller than the mirror itself. The masked mirror should display a good edge. If the outer edge of the extrafocal disk looks different with the stop than it does at full aperture, this confirms the diagnosis. Go back to figuring, with more attention given to the edge!

Rings and More Rings.   Carefully examine the diffraction rings inside the defocused star image. Can you see them at all? In a telescope with a severely lumpy mirror—either zoned or dog-biscuited—the rings may be hard to see, or invisible. Lumpiness introduces random path-length errors that obliterate the regular ring pattern of extrafocal diffraction. For observing with low magnification, lumpiness is not harmful, but the mirror will perform poorly on the planets.

Zoning is common in fast, heavily figured mirrors of large aperture. Zones tend to be obscured in the Foucault test on short-focus mirrors because the paraboloidal shadows are so strong, but in long-focus

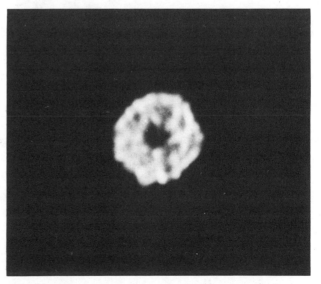

The out-of-focus image of a star distorted by atmospheric turbulence.

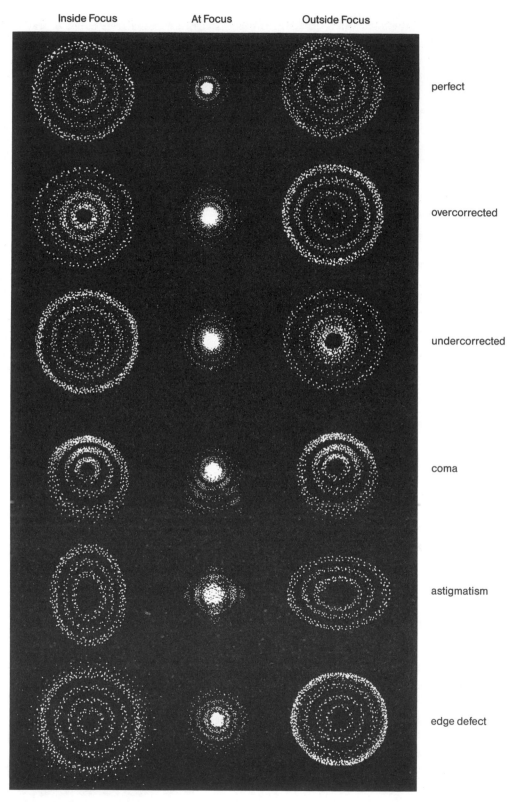

Inside Focus   At Focus   Outside Focus

perfect

overcorrected

undercorrected

coma

astigmatism

edge defect

Perfect and imperfect optical system shown inside, at, and outside focus for star-testing. Note that imperfections are difficult to distinguish at focus but readily diagnosed by checking the defocused image.

mirrors, you'll see zones as narrow, concentric circles on the mirror's surface. You should not expect heavy zoning in a 6″ f/8 mirror. Severe dog-biscuiting shows up as reduced extrafocal diffraction-ring contrast, but without direct comparison to other telescopes, it is difficult for a beginner to detect. Dog-biscuiting is visible in the Foucault test as randomly distributed hills and valleys on the mirror.

Examine the distribution of light across a star image, both inside and outside focus, defocused so it shows five or six rings. The disks should be identical. If you see a zone that is bright on one side of focus and dim on the other, the mirror may have one or more broad, gentle figure errors. (If the intensity of the entire extrafocal disk varies from edge to center, that's spherical aberration, which we'll discuss next.) Broad, zonal-figure errors differ from severe zoning in being relatively wide and smooth, with gentle slopes. Mild dog-biscuiting is a common ailment in homemade mirrors; luckily, it is really not much of a problem.

For comparison, use a small refractor made by a reputable manufacturer; carefully examine the images it forms. Refractors generally have very smooth, zone-free surfaces, so the extrafocal ring pattern should appear almost perfect.

Checking the Paraboloid. How good is the overall figure? If the mirror has no other major problems, departure from a perfect paraboloid will show up as spherical aberration. When the mirror is undercorrected—that is, if its figure lies somewhere between a sphere and a paraboloid—the focal length will be too long in the inner zones and too short in the outer zones. When the mirror is overcorrected, with a figure stronger than a paraboloid, the outer zones will focus long and the inner zones short.

Defocus the star until you see a disk with five or six diffraction rings. Go from inside focus to outside, then back and forth several times. Compare carefully. *If the mirror is uncorrected when it's inside focus, the outer edge of the extrafocal disk is too bright; when it's outside focus, the center of the extrafocal disk is too bright.* For an overcorrected mirror, the reverse is true: When you view the star image inside focus, the center is too bright; when it's outside focus, the edge is too bright. If the extrafocal disks look identical with six rings, reexamine the diffraction patterns defocused to two or three rings.

A word of warning: This test is quite capable of turning up errors that won't make any real difference in how well the telescope performs normal observing tasks. Try to be critical but not overcritical.

If you star-test regularly while figuring, don't miss the opportunity to examine star images when the Foucault test shows that your mirror is a good sphere. It will then have a bit more than ¼-wave of pure spherical aberration if it's a 6″ f/8. This amount of spherical aberration degrades the image quality, but only a bit more than would be acceptable if you weren't too fussy (or didn't know any better).

Star-Testing a Refractor. Begin by collimating. With no eyepiece in place, shine a penlight down the focuser tube. Align the objective so that the reflections of the penlight from the lens surfaces are exactly centered in the lens. This method provides all the collimation it'll probably ever need.

Some achromatic objectives show coma when misaligned; others are coma-corrected, showing astigmatism instead. (In a misaligned paraboloidal mirror, both coma and astigmatism are present, but coma masks the astigmatism.) If you see astigmatism, it may be due to misalignment rather than to figure error.

Lumpiness is seldom a problem with refractors, but check for inhomogeneity in the glass. This appears as irregular wisps, streaks, or distortions crossing the intra- and extrafocal pattern. In a modern lens, with optical glass as good as it now is, this is an inexcusable defect. If you have an "old-time" objective, remember that you also have the glass technology of a bygone era.

Next, examine the lens for spherical and chromatic aberration. Remember that the focal length of an achromat varies with color. The shortest focal length and best overall correction should occur in yellow-green light. To check spherical aberration then, place a deep-yellow or green filter over the eyepiece and inspect the image for spherical aberration. A small amount doesn't matter, but ideally there should be none.

Remove the filter and slowly roll through focus. Inside focus, you should see a clearly defined extra-focal disk with concentric diffraction rings enveloped in a blue-purple haze. As you focus outward, the rings will move toward the center and disappear. At best focus, the Airy disk will have a yellowish tinge and be surrounded by a purple halo. On a night with good seeing, several diffraction rings should be visible.

Just past focus, the Airy disk will expand into a yellow-green ring; then a delicate red spot will form—this is the focal point of red starlight. Slightly beyond the red focus, blue and violet rays reach a blurry focus. Farther outside focus, a haze of blue light overlying the extrafocal image may partially obscure diffraction rings. When you're checking for spherical aberration, a deep yellow green filter removes this unfocused light.

While doing these tests, check for signs of decentered or wedged lens elements. Focus on a star high in the sky; carefully examine it for asymmetric color effects or asymmetric extrafocal patterns. Improper spacing may cause spherical aberration, although spherical aberration may also result from poor design or errors in the construction of that particular lens. It may be

The Foucault test permits determining a mirror's figure in the shop—without testing on stars.

necessary to remount an objective, especially an "unknown" or secondhand lens, several times before it performs right. If you find these errors in a ready-made and assembled lens, return it as defective.

An apochromatic lens should be nearly totally free of both spherical and chromatic aberration. Depending on the design of the lens, you may see a tiny blue or red dot just outside best focus when the blue or red rays focus, but aside from this trivial defect, any image less than "textbook perfect" is unacceptable.

TESTING IN THE SHOP

During polishing and figuring, you'll need to know the effect of a session of figuring. Since it is neither efficient nor practical to star-test at the end of every session, an indoor test is needed.

The drawback of this test is that it requires a Foucault tester, but this is not difficult to construct. The Foucault test is quite sensitive to the smoothness of the mirror surface, and although the test "shadows" take some effort to understand, you'll soon get a feel for them.

Just what *is* the Foucault test? If a beam of light from a bright source located at the mirror's radius of curvature shines on the mirror, the mirror will form an image of the source. If you then block part of the image with a knife edge and look at the mirror with light that has gone past the knife, you will see the mirror's "defects" in high relief.

To understand how it works, first consider a spherical mirror. If a narrow, brightly illuminated slit lies exactly at the mirror's center of curvature, light will strike the mirror perpendicular to its surface at all points and be reflected from it at the same angle, returning to focus near the source. If you place your eye in the returning light, the mirror surface appears to be illuminated.

Geometry predicts that all returning rays come to focus at a single point. Suppose that you now block some of the light coming to focus with a sharp metal knife blade—what will you see? If the blade is in front of or behind the focal point, you'll see its "shadow" as it blocks some of the light from one side of the mirror while the other side still appears fully bright. On the other hand, if you cut the light beam exactly at the focal point, light from every point on the mirror

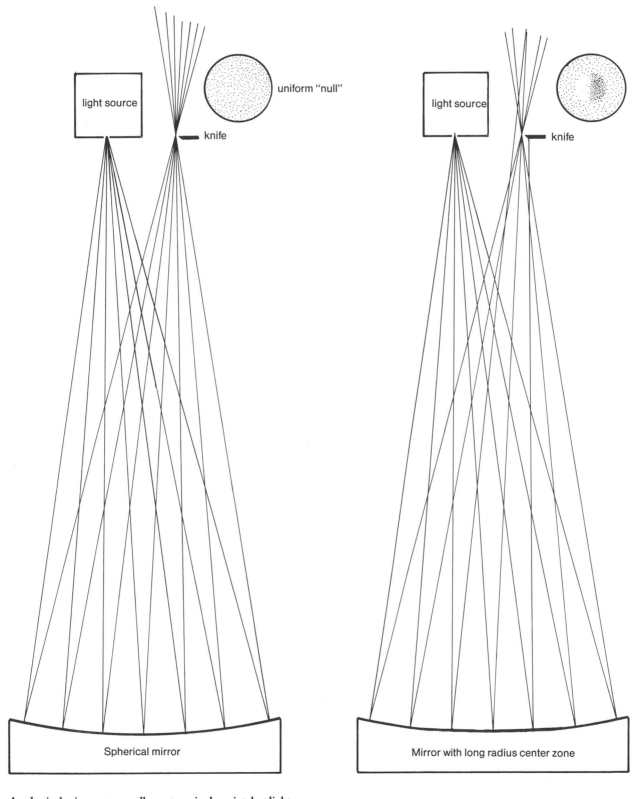

light source

uniform "null"

knife

light source

knife

Spherical mirror

Mirror with long radius center zone

A spherical mirror returns all rays to a single point, but light reflected from slight imperfections passes the knife edge.

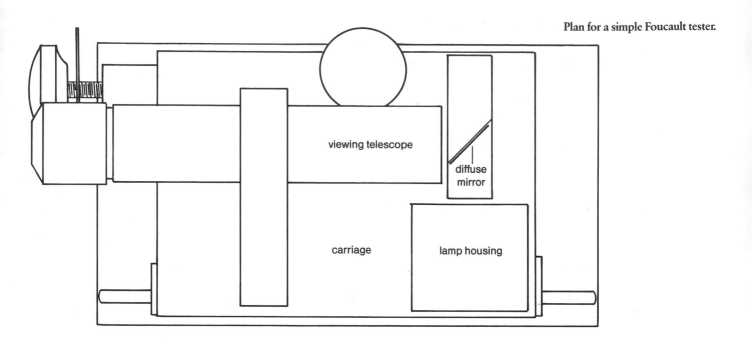

Plan for a simple Foucault tester.

viewing telescope

diffuse mirror

carriage

lamp housing

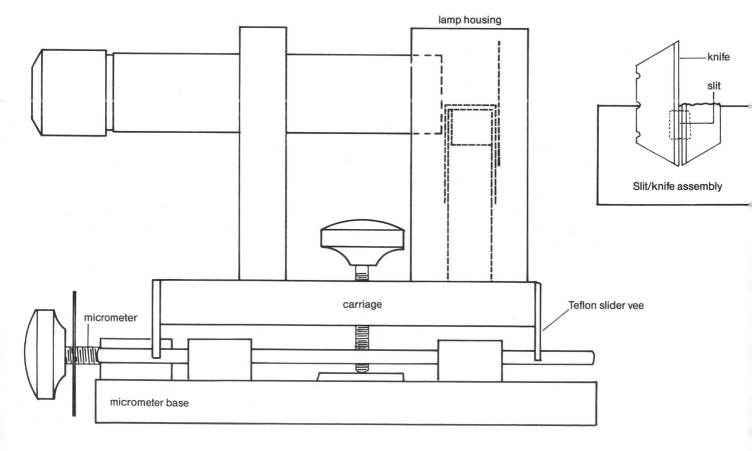

lamp housing

knife

slit

Slit/knife assembly

micrometer

carriage

Teflon slider vee

micrometer base

will be cut off simultaneously and the mirror will appear to darken uniformly. A spherical mirror, therefore, gives a "null."

Now suppose the mirror is a somewhat imperfect sphere, one with a slightly long radius in the center. Now what will you see? With the knife edge at the focus of the outer zones, the zones will dim uniformly as they did before. However, the knife is now inside the focal point of the inner zones, so you see a "shadow" on one side of the zones and full brightness on the other. The defective mirror resembles a flat, gray surface with a shadowed hill on it. If you move the blade farther from the mirror so that it cuts the focus of the inner zones, the zones will dim uniformly and the shadow/bright pattern will appear in the outer zones. The center of the mirror will then look like a flat, gray spot at the bottom of a shadowed depression.

Thus, whatever zone of the mirror has a radius of curvature matching the precise distance between the mirror and the tester will appear as a gray nulled zone; other regions will appear as shadowed hills or holes. The Foucault test offers the chance to survey the radius of curvature of each zone of the mirror. When the test shows that every part of the mirror has the same radius, for example, you have a sphere. But what of the paraboloid? The answer—and the power of the test—lies in the test's ability to break the smooth paraboloidal surface into a series of "almost-fitting" spheres and to explore whether or not the observed radii of curvature match theoretical values.

CONSTRUCTING A TESTER

While Foucault testers take many forms, the tester described here is relatively easy to build, easy to use, and works accurately and reliably. It consists of a lamp in a metal housing, a slit-and-analyzer assembly, a viewing telescope mounted on a carriage that slides smoothly back and forth with them, and a micrometer base, which measures the position of the carriage. With the exception of Teflon plastic, the materials for it are readily available in hardware stores everywhere, and the tester can be built entirely with hand tools. Since you will need Teflon for the telescope's mounting, the Teflon you need for a tester can be purchased

at the same time (see Appendix A). If you can't find Teflon, substitute acrylic or nylon plastic even though they won't slide as smoothly.

Begin by obtaining the optical parts: a bright, compact light bulb (a 110-volt, 25-watt clear or frosted appliance lamp works well); a small achromatic lens 15mm to 30mm in diameter with a focal length between 50mm and 80mm (Edmund Scientific is a good source); and a 25mm to 30mm focal-length eyepiece in a 1¼″ barrel (use the low-powered eyepiece intended for the telescope). Buy two hardware-knife blades for the knife edge and slit. You'll also need a socket for the lamp, an unpainted metal (*not* plastic) electronics project box roughly 2″x2″x4″, a toggle switch, a grounded cord and plug, a 1¼″ sink-trap tube (from the plumbing department), and a small amount of 1/32″ sheet aluminum or steel.

Construct the lamp housing first. This metal box holds all the electrical parts. Designate one small end of the box as the bottom, one side as the outward-facing side, and the side opposite that as the inside. Mount the lamp and its socket in the box with the filament roughly 2¾″ from the bottom. Mount the switch and cord near the bottom of the box on the outside side. Ground the housing to the green (ground) wire of the cord assembly.

Drill a ½″ hole on the "inside" opposite the lamp filament; the light exiting this hole illuminates the slit/analyzer assembly. An optional ¼″ hole on the "front" side will make alignment easier. Once you see exactly how the lamp box will fit into the test unit, drill ¼″ holes around the bottom and top ends for ventilation, but plan carefully so that light escaping from them won't shine in your eyes!

Next, assemble the viewing telescope. Saw the large end of the drain tube with a T cut, then bend the tabs so the tube accepts a 1¼″ eyepiece snugly. Measure the distance between the achromatic lens and the eyepiece when the combination is focused 6′ to 10′ away; then cut the drain tube to this length. Mount the lens to the front end of the tube with glue or tape, slip the eyepiece in, then test this little telescope's performance. It should give a sharp image with a magnification between 2x and 3x, and it must be focusable (by sliding the eyepiece) from 4′ to infinity.

With these elements completed, the next job is constructing the carriage and micrometer base. These are both made from 1″ lumber. The carriage is a block roughly 4½″ wide by 6″ long; the base is the same width by 8″ long. Also cut out two 1″x1″ blocks; two 2″x2″ blocks from the same material. Obtain two knobs (try a radio-parts store if you can't find these in your hardware store); two pieces of ⅛″ Teflon 1¼″ square, one piece 1″x2″, and another piece 1¼″x½″; an 8″ length of steel rod ¼″-diameter (it's sold in 36″ lengths); two thin, weak springs each about 3″ long; three ¼x20 tee nuts; four small eye hooks; assorted wood screws; a 4″x¼-20 carriage bolt, and a 2″x¼-20 carriage bolt.

Carriage bolts have round heads, but they may be slightly eccentric. Chuck each bolt in a drill, then press the head against a sheet of fine emery cloth backed with a thick cloth pad. Run the drill at high speed for a minute or two while swiveling it about the head of the bolt. The bolt should become smooth, shiny, and perfectly round.

Drill a 5/16″ hole down the length of one of the 2″x2″ wooden blocks. Glue one tee nut at one end of the hole and place another at the other end, but do not glue it. Let the glue dry. Thread the 4″ carriage bolt through both tee nuts. The bolt should turn smoothly with very little wobbling. If it binds, turn the second tee nut until the motion of the bolt is smooth; then glue the tee nut in place. This is the micrometer assembly. Nail and glue it to one end of the base block. Drill a ¼″ hole through each of the 1″x1″ blocks, then push the ¼″ rod through them. Slide the blocks 2″ in from each end; then nail and glue them to the base block.

Glue and nail the other 2″x2″ block to the underside of the carriage 2½″ from the back end. The head of the micrometer bolt bears against this block. Tack the smallest piece of Teflon at that spot. Cut V notches in the two 1¼″ pieces of Teflon in mirror-image positions; then attach the Teflon with wood screws to the ends of the carriage block. Trim the Vs until the carriage rides level on the steel rod.

At the middle of the opposite side of the carriage block, drill a 5/16″ hole and install a tee nut in it. Thread the short carriage bolt through it and screw the knob on the upper end of it. Turning this bolt will move the knife edge laterally. On the base block under the bolt, tack the remaining piece of Teflon. Screw in the screw eyes, two in the base and two on

the underside of the carriage, and attach the springs. Experiment, moving the screw eyes or changing the springs for weaker or stronger ones, until the carriage block sits snug against the head of the micrometer bolt but not so strongly that it hops off the metal rod.

To complete the carriage, install a micrometer dial, a 2″-diameter cardboard or plastic circle, on the end of the carriage bolt next to the threaded knob. Since each turn of the bolt advances the carriage by 0.050″ mark the dial off into 50 parts—ten major divisions with five subdivisions each. Label the major divisions 00 through 45 by 5s so that when you turn the bolt, the reading *increases* when the carriage moves *away from* the mirror.

Next, complete the optical parts of the tester. Mount the lamp housing on the carriage with wood screws, then position the viewing telescope beside it roughly 1/16″ away. Cut a holder for the telescope. Design it to grip the telescope tightly, yet not to interfere with the lateral adjustment knob.

The final job is making the heart of the tester—the slit/knife edge assembly—from a piece of thin aluminum approximately 2½″ x 4″ that slips over a vertical block of wood roughly 2″ wide by 3″ tall. Make a 45° cut in the upper corner, centered on the height of the lamp-housing hole. Screw this piece to the carriage so the center of the cut lies at the intersection of the center of the lamp-housing hole and the optical axis of the viewing telescope. Glue a piece of aluminum (burnished with fine emery cloth) to the 45° cut: This "mirror" serves to direct light to the slit while also diffusing it.

Cut and bend a three-sided aluminum piece to fit over the wooden block. Determine its precise dimensions from the tester. When slipped over the block, it should catch all the light coming from the large hole in the lamp housing. About 1″ from one end, drill a line of small holes in it, then file them into a slit ⅛″ wide by ½″ high in line with the viewing telescope. The slit must lie on the optical axis of the viewing telescope. Glue a hardware-knife blade overlapping this hole with half of the blade's length protruding above the top of the U. Break another hardware-knife blade in half. This blade is brittle, so clamp it in a vise and snap off the other end with a quick hammer blow. Wear safety goggles!

Temporarily tape the half-blade to the slit assembly

The Foucault tester can measure errors of less than a millionth of an inch on the surface of a mirror.

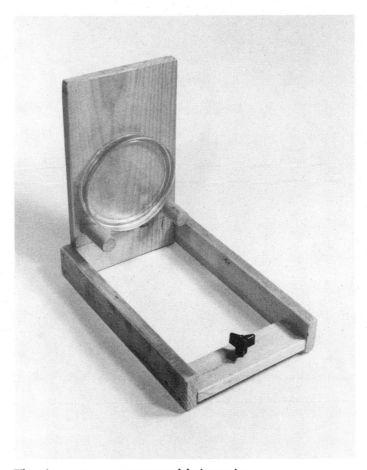

The mirror rests on a support stand during testing.

(use masking tape) opposite the lower end of the intact blade. Slip this assembly over the wood block, turn on the lamp, and watch the slit light up. Unstick the tape holding the half-blade and move it closer until only a uniform, bright band of light a few thousandths of an inch in width shines through.

If necessary, alter the positions of the slit/analyzer assembly and telescope so that light from the mirror that passes the knife edge reaches the viewing telescope. If the first slit/analyzer assembly is unsatisfactory, make another one to replace it.

Finally, construct a mirror stand to hold the mirror for testing. This is a simple device: a board about 7″ wide and 12″ tall with two ¾″ dowels 4″ apart protruding from it. This board is held upright by two 1″x3″ boards 12″ long, joined at the front by another 1″x3″ with an "elevation" knob (similar to the Foucault

tester's lateral adjustment) for aligning the mirror when it's tested. With the completion of the tester and mirror stand, you are ready for figuring.

OBSERVING TEST SHADOWS

Making the Foucault test is simple; interpreting the results is far more difficult. Nonetheless, always keep in mind that the test measures *differences in the radius of curvature* of the mirror—not the hills or holes the shadows resemble—and you'll stay on track.

Begin the test session by finding the approximate radius of curvature; then slide the carriage toward the mirror three or four turns of the mircometer knob. Advance the knife across the slit image, noting the direction of travel. This illustrates the First Law of Foucault Testing: *Inside* the radius of curvature, the

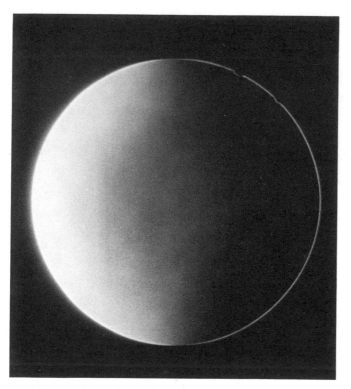

The appearance of the Foucault shadows of a good 6″ f/8 paraboloid. Central null

30% zone null

shadow advances in the *same* direction as the knife edge travels. Because the telescope reverses the image and the lateral motion knob rotates rather than pushes, it's easy to get mixed up, so reformulate the law in terms you see and feel; for example, *inside* the radius of curvature, the shadow advances in the *same* direction as the front side of the knob turns.

The Second Law of Foucault Testing is the complement of the First Law: *Outside* focus, the shadow advances in the *opposite* direction as the knife edge travels. Run the micrometer back outside focus to verify it. With these rules firmly implanted, you're prepared to test. (Caution: Since the viewing telescope inverts and reverses the image of the mirror, don't become confused by other books on mirror testing. Left and right may be reversed from the way you see them through your tester.)

Advance the carriage micrometer knob and diddle the transverse knob as the carriage moves toward focus. The shadow will move increasingly fast across the mirror, and if you advance the carriage farther, the

shadow will finally begin moving in the opposite direction. Backtrack until the shadows flicker across the mirror with the most delicate touch.

If the whole mirror grays and blanks out uniformly, it has a spherical figure. If part of it blanks out and other zones appear as bright and dark, the carriage is at the radius of curvature of the nulled zone. The other zones—those that look bright and dark—have either shorter or longer radii. Note which side the shadow comes from, then apply the laws of Foucault testing. Is the carriage inside or outside the radius of curvature? Test the answer. Move the carriage either inward or outward until that zone nulls.

To aid in visualizing the surface, lightly sketch a straight line representing the mirror's surface. Darken the line to indicate the zone that is at null. Sketch the other zones relative to that line—shorter radii as arcs of a concave curve and longer radii as convex sections—to create a cross section of the surface. To decide, note the direction the shadow comes from in that zone—don't rely on appearances. Remember that the

70% zone null

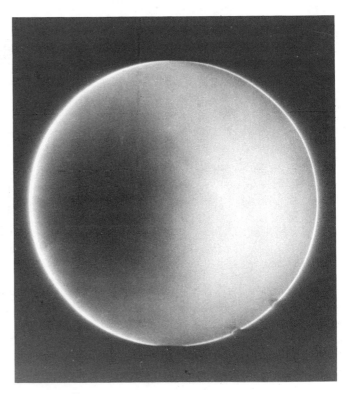

85% zone null

appearance of any section depends on which zone happens to be the reference sphere. Shift the knife edge longitudinally to null another zone, then chart it too. Look for the relation between the sections in terms of radius of curvature.

While you're in the early stages of figuring and using cross sections to guide your next step, try to find a reference sphere—either directly with the tester or indirectly by mentally "bending" cross sections—which makes figure errors appear as convex or regions of too-long radii. Plan the figuring strokes to polish away these hills.

What does a paraboloid look like? If we imagine it as a series of spherical zones, a paraboloid is a surface whose radius of curvature increases with the radius on the mirror. How it looks under testing, of course, depends on where our reference sphere lies. If the reference zone is the center of the mirror, all zones outside the center have longer radii and are therefore convex. The mirror's profile resembles a gently rounded swelling.

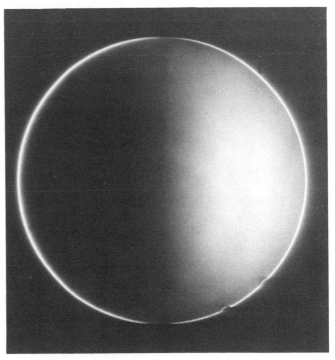

Edge null

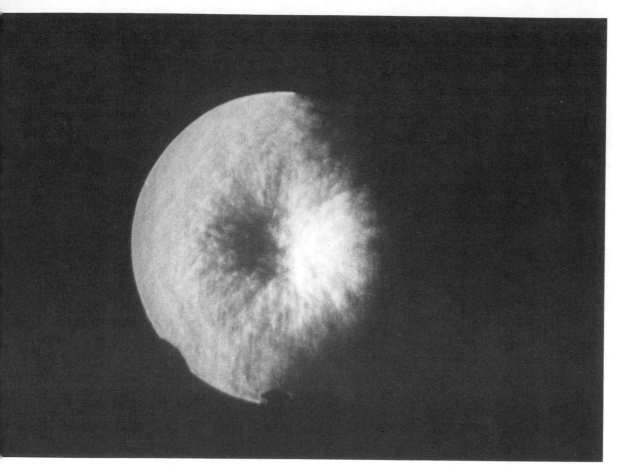

This mirror has a deep central hole and many small, deep errors—probably caused by rapid strokes and heavy pressure during figuring.

A bad case of "dog biscuit" combined with deep gouges made by a defective facet in the polishing lap.

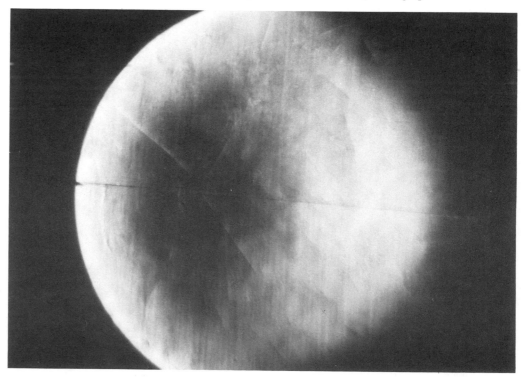

Suppose you move the carriage farther from the mirror. Zones inside the null have shorter radii, so appear concave. Zones outside the null have longer radii and still appear convex. The half-profile goes up, then down again. The whole mirror has the famous "doughnut profile" of mirror-testing literature!

If you back the carriage farther from the mirror until the outer zones of the mirror null, you'll see that all the central zones have shorter radii than the nulled zone, appearing concave. The mirror profile now resembles a broad, shallow soup dish.

All three appearances are indicative of a paraboloid-like figure, but they are not *unique* to the paraboloid. All conic sections—ellipses, hyperboloids, oblate spheroids, and paraboloids—share this shadow pattern. The only difference is the longitudinal spread of radii. Spheres are the degenerate case: All zones are equal.

Ellipses have a smaller range of radii than paraboloids; hyerboloids have a larger range. For oblate spheroids, the order of radii is reversed but the patterns look the same. To finish figuring your mirror, appearances are not enough; you must measure the range of radii of curvature using the micrometer built into your Foucault tester.

ZONAL FOUCAULT TESTING

Foucault test shadows are qualitative in nature. They are good for judging smoothness and blending but nearly useless for determining *how much* correction we have given the mirror. Zonal readings and the Foucault graph make the test quantitative.

A paraboloidal surface can be represented as a series of spherical surfaces of progressively longer radius.

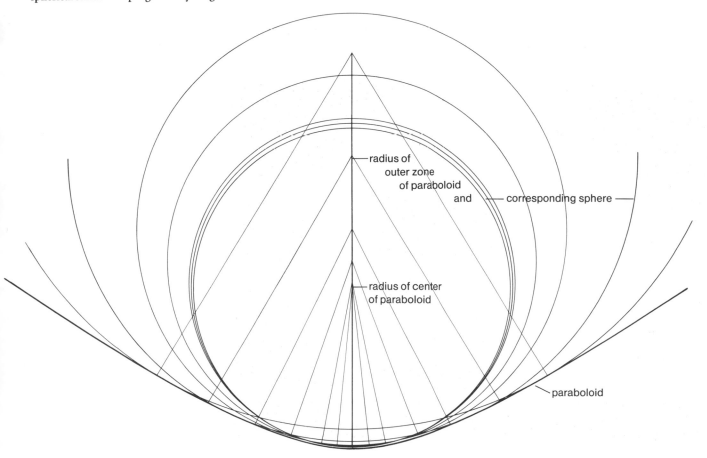

radius of outer zone of paraboloid and — corresponding sphere —

radius of center of paraboloid

paraboloid

To map the surface, the test requires the radius of each of the succession of spheres nulled at focus. We then compare these measured radii to the radii expected from the paraboloid's geometry. For a given radius, r, on a paraboloidal mirror, the radius of curvature, R_{zone}, that best fits that zone is:

$$R_{zone} = R_0 + r^2/2R$$

R_0 is the radius of curvature of the center of the mirror. The tester is already a distance R_0 from the mirror and since we will need to know the difference between R_{zone} and R_0, we define ΔR as the small change in the radius we're measuring:

$$\Delta R = r^2/2R$$

(Note: the treatment in this book differs from the traditional one in using the change of radius of curvature rather than the displacement of the knife edge. The slit and knife move together on the tester described in this book; thus it measures ΔR directly. Other books derive the movement of the knife edge relative to a stationary light source; the knife edge displacement is then twice ΔR.)

How does the difference in the radius of curvature vary with the zone on the mirror? Examine the table below for a 6″ f/8 mirror. Remember that the radius is twice the focal length, or 96″.

Radius	$r^2/2R$	R_{zone}
0.0	0.0	96.0000
0.5	0.0013	96.0013
1.0	0.0052	96.0052
1.5	0.0117	96.0117
2.0	0.0208	96.0208
2.5	0.0326	96.0326
3.0	0.0468	96.0468

Note that this change of radius is quite small. This is why the carriage of the tester measures small longitudinal displacements. You will measure, one

Two masks used to divide the mirror surface into five zones for testing.

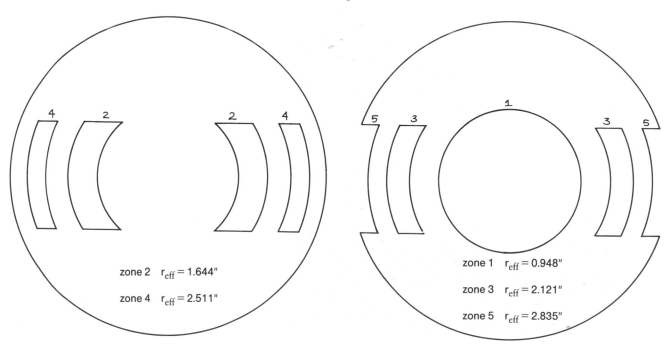

zone 2 r_{eff} = 1.644″

zone 4 r_{eff} = 2.511″

zone 1 r_{eff} = 0.948″

zone 3 r_{eff} = 2.121″

zone 5 r_{eff} = 2.835″

zone at a time, the actual value of ΔR for your mirror.

How can you best measure the radius of one zone? The most practical way is with a paper mask used to isolate just that part of the mirror.

Construct zonal masks from light, stiff cardboard—file-folder material is perfect. Measure the diameter of the mirror, divide it by 2 to obtain the radius, then multiply the radii tabulated below to obtain radii for cutting the masks' zones. (Don't just assume that the radius of a 6″ mirror is 3.00″. Measure it. More likely than not, you'll find it's something like 2.87″ or 2.96″. Multiply the radius times $r_{in/out}$ to get the radius of the zone for the mask. The value r_{eff} is the radius of the center of each zone—you'll need it later.)

Draw the mask patterns carefully with an accurate compass; then cut the openings in the masks with a sharp Exacto knife.

How many zones? Start with three zones, then progress to five. Stay with this number until you feel thoroughly comfortable with zonal Foucault testing. For large, fast mirrors, or for the final tests of a mirror you are confident you have finished, use seven test zones. Here are sample tables for three-, five-, and seven-zone masks:

FIVE-ZONE MASK

Zone	$r_{in/out}$	r_{eff}
	.000	
1		.316
	.447	
2		.548
	.632	
3		.707
	.775	
4		.837
	.894	
5		.945
	1.000	

THREE-ZONE MASK

Zone	$r_{in/out}$	r_{eff}
	.000	
1		.408
	.577	
2		.707
	.816	
3		.913
	1.000	

SEVEN-ZONE MASK

Zone	$r_{in/out}$	r_{eff}
	.000	
1		.267
	.378	
2		.463
	.535	
3		.598
	.655	
4		.707
	.756	
5		.802
	.845	
6		.886
	.926	
7		.964
	1.000	

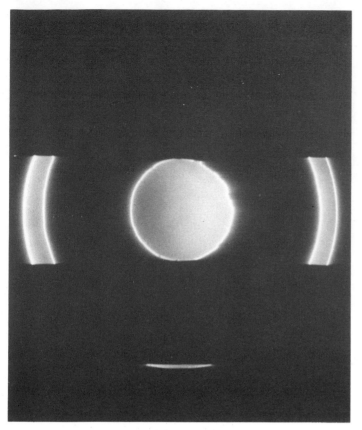

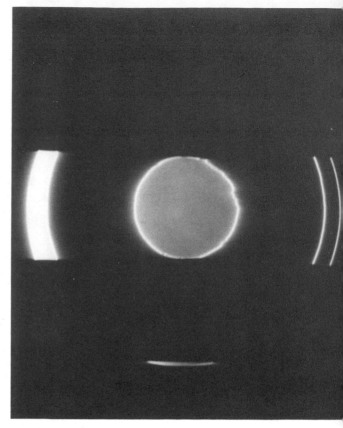

Edge zones balanced

Center zone balanced

Zonal testing consists of finding the radius of curvature at which a pair of zones "balances" in brightness.

Note that the middle zones in every case has an effective radius of .707—the so-called "70-percent zone" of amateur telescope-making literature. Equal areas of the mirror lie inside and outside it.

Whether you measure three, five, or seven zones, each zone in these masks represents the same fraction of the mirror's area and contributes an equal amount of light to the final image. The zones bunch up toward the outer edge of the mirror, emphasizing the relative importance of the edge zones as compared to the center.

To make zonal readings, set up and align the tester and mirror as for examining the Foucault shadows, then allow everything to reach thermal equilibrium. After a polishing session, allow an hour for the glass to reach air temperature. Check to see that air currents are not disturbing the shadows. Place the first mask in front of the mirror, then look through the viewing telescope. Each hole in the mask will be outlined in diffracted light, with some zones bright and some zones dim.

Cut the slit image with the knife while watching

two opposite mask openings. Remember now, you are looking at an isolated section of a sphere, trying to locate the null. If the shadow travels in the same direction as the knife, the knife is inside its radius of curvature, so move the carriage back; if the shadow moves in the opposite direction as the knife, move the carriage inward. At some distance from the mirror, both zonal openings in the mask will darken simultaneously—read the scale and write down the reading. Repeat this procedure for each zone; then run through the whole sequence two more times in different order.

Zonal reading is slightly maddening. The judgments are delicate and difficult. The slightest touch on the lateral knob makes all the shadows seen through the mask flicker, but your job is to exactly balance the dimming of two sides of the zone by moving the carriage. "Where is that balance point?" you'll wonder. If it seems hard to locate, move the carriage longitudinally half a turn of the micrometer knob forward or backward. The imbalance will then become obvious. It may help to bracket the balance point by trying to locate equal imbalances on either side, then closing in on them.

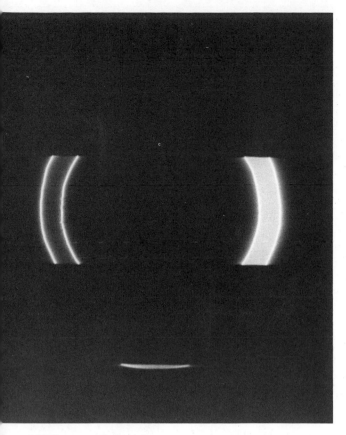

Middle zone unbalanced

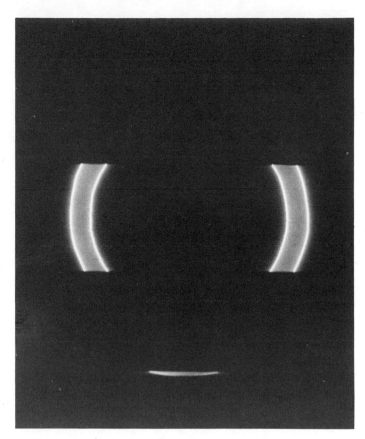

Middle zone balanced

Below is a sample set of readings made for seven zones on three masks for a 6″ f/8 mirror. Mask A has openings for zones 1, 4, and 7; mask B for zones 3 and 6; and mask C for zones 2 and 5. Readings are recorded in thousandths of an inch, and the sequence runs from left to right as each mask was used in its turn.

The distance between the mirror and the tester must remain constant to within 0.001″ during the entire set, so be careful not to bump the mirror when switching masks. Don't lean on the table supporting the tester; don't bump the viewing telescope with your nose. Don't, in fact, touch anything on the tester except the longitudinal and lateral adjusting knobs!

Don't be tempted to cheat either. Cover the record sheet as you make each set of readings. Don't look at the numbers; don't try to come back to the same settings. There are limits to how closely zonal readings can be made and the scatter serves as a reminder of the ultimate accuracy of the test.

MASK

	A	B	C	B	A	C	C	A	B
Zone									
1	23				28			25	
2			37			31	34		
3		44		38					46
4	49				51			50	
5			55			62	57		
6		59		58					63
7	75				75			72	

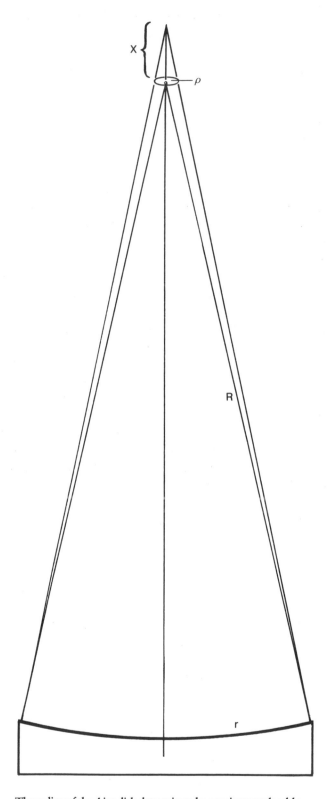

The radius of the Airy disk determines the maximum tolerable error in the radius of curvature of any given zone.

THE FOUCAULT GRAPH

A graph is the quickest and easiest way of seeing how well the measured zonal radii compare with the expected readings. A Foucault graph consists of a plot of the effective zonal radius, r_{eff} (see the tables above used to make masks), against the variation in its radius of curvature, ΔR. Since you don't *know* the precise value of R, the values of ΔR "float." To accommodate this, plot the expected values and "tolerance envelope" on one sheet of paper; plot the measured values on a transparent overlay; then slide the overlay with the measurements on it up and down. If all the measurements fit within the tolerance envelope, the mirror is acceptable. If they do not, you now have some idea of the mirror's figure errors for the next round of figuring.

What is this "tolerance envelope"? Recall our table of ΔR versus zonal radius. How much error in ΔR is acceptable? Suppose that the radius of a zone is different from ΔR by an amount X. If we adopt as our tolerance criterion that the light must fall within the stellar diffraction disk, we should obtain sensibly perfect performance. The radius of the Airy disk, ρ, at the focus of a telescope focused on a star is:

$$\rho = 1.22\lambda F/D$$

where F is the focal length, D is the aperture of the telescope, and λ is the wavelength of yellow-green light to which our eyes are the most sensitive: 0.550 micrometers, or 21.6 millionths of an inch. Evaluating this equation for a 6″ f/8 mirror:

$$\rho = .00021''$$

The triangle formed by the radius tolerance X, ρ and the defocused light ray is similar to the triangle this ray forms with R and r_{eff}. This is true for both inside and outside focus. We write the following equality involving the tolerance X:

$$X/2\rho = R/r$$

Solving for the Foucault tolerance:

$$X = 2\rho R/r$$

Thus for any zone, r_{eff}, the minimum acceptable zonal reading is:

$$\Delta R_{min} = \Delta R - X$$

and the maximum acceptable zonal reading:

$$\Delta R_{max} = \Delta R + X$$

Tabulating them for a 6″ f/8 mirror is the table above right.

We may also adopt a more demanding criterion, one requiring that zonal error not deflect any ray outside the central *half* of the Airy disk. This criterion demands that rays will fall within the central 25 percent of the area of the diffraction disk. To satisfy this criterion, X becomes X/2, and we can construct the following table below right.

Plot these values on a graph. Do ΔR first: It rises steadily with increasing zonal radius; in fact, it's a parabolic curve! The tolerance curves form a hornlike envelope on either side of the ΔR curve, tighter toward the edge of the mirror and more open toward the center. The most difficult zones to figure and make zonal readings for are also the most important.

The lower the f/ratio of the mirror, the more rapidly ΔR rises and the tighter the tolerance envelopes cling to it. To compare the relative ease of figuring a 6″ f/8 and a 6″ f/4 mirror, calculate and plot the ΔR and tolerance curves for each. Try plotting these curves for 12″ f/4 and f/8 mirrors. You will then understand why a 6″ f/8 mirror is suitable for novice mirror-makers.

Plot the zonal measurements. Draw the y-axis on a sheet of clear plastic or tracing paper, then plot all three sets of readings. Don't average them or discard "faulty" readings—you are after the truth about the mirror. Slide the overlay sheet up and down until the readings best fit the theoretical curve. (Note: The y-axes must overlay each other; the only permissible shift is up and down, in effect, adjusting the value of R.)

If all or most of your readings fit within the tolerance envelopes, the paraboloid is satisfactory. As a rough rule of thumb, the easy tolerance envelopes represent ¼-wave wavefront error and the demanding tolerance approximately ⅛-wave wavefront error. The scatter in the zonal readings gives some indica-

ZONAL TOLERANCES—"EASY" CRITERION

Radius	ΔR	X	ΔR_{min}	ΔR_{max}
0.0	0.0000	——	——	——
0.5	0.0013	0.0814	−.0801	0.0827
1.0	0.0052	0.0406	−.0354	0.0458
1.5	0.0117	0.0271	−.0154	0.0388
2.0	0.0208	0.0203	0.0005	0.0411
2.5	0.0326	0.0162	0.0164	0.0490
3.0	0.0468	0.0136	0.0332	0.0604

ZONAL TOLERANCES—"DEMANDING" CRITERION

Radius	ΔR	X/2	ΔR_{min}	ΔR_{max}
0.0	0.0000	——	——	——
0.5	0.0013	0.0407	−.0394	0.0420
1.0	0.0052	0.0203	−.0151	0.0255
1.5	0.0117	0.0131	−.0014	0.0248
2.0	0.0208	0.0102	0.0106	0.0310
2.5	0.0326	0.0081	0.0245	0.0407
3.0	0.0468	0.0068	0.0400	0.0536

tion of the accuracy of the Foucault readings on the Foucault graph.

All conic sections graph as parabolic curves, with the same form as a paraboloid, but a different magnitude of ΔR. Estimate the percentage of parabolization from the curve. If the mirror is undercorrected, the center zones are "high" relative to the edge zones. The radius of the central part of that mirror is too short. Additional work on the center will shorten the radius of this part of the mirror.

When using Foucault graphs as feedback for the next polishing session, a quick way to gain a feel for where to polish is to slide the plotted points upward until the lowest ones are tangent to the theoretical curve. The high zones are those areas that lie above the curve. In the case of a partially parabolized (ellipsoidal) mirror, the outer zones will match and the inner zones will be high; continue the parabolizing stroke to deepen the center.

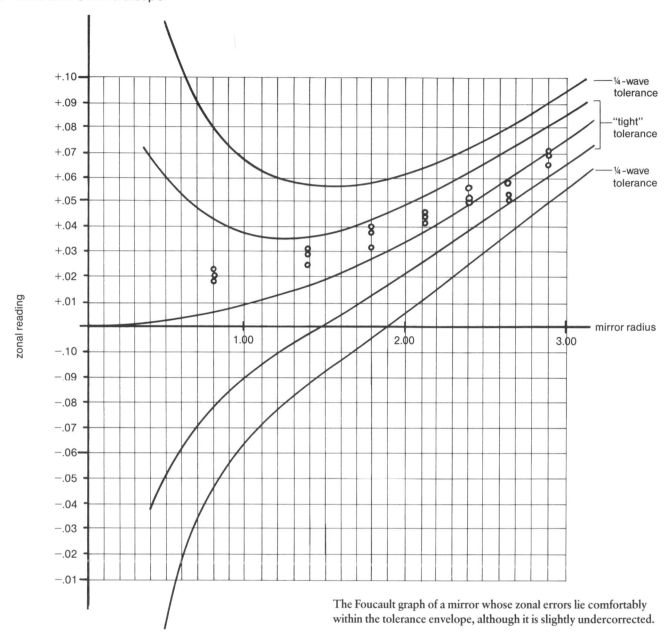

The Foucault graph of a mirror whose zonal errors lie comfortably within the tolerance envelope, although it is slightly undercorrected.

The strength of the graph lies in its ability to map out the *broad* features of the figure, the overall state of correction. Generally, the Foucault graph is not helpful in evaluating *small* features because the three-, five-, or seven-point curve is too coarse to show them. For figure smoothness, rely on your eye's estimate of the "blendedness" of the shadows. Harsh features simply shouldn't be there.

Finally, run a star test from time to time. If you find yourself wrestling night after night in an attempt to get a perfect graph, but the star test shows a near-perfect star image, it's time to remind yourself that you're trying to construct a telescope—not a graph!

IN SEARCH OF OPTICAL PERFECTION

No telescope is perfect. Why, then, do we even talk about optical perfection? We have seen that a perfect optical system is limited, ultimately, by the wave nature of light. At some level of perfection, there will be no gain from further improving the surface.

Lord Rayleigh, in the 1800s, suggested a simple rule of thumb for optical quality: If light coming to a focus has no optical path difference error exceeding one-fourth the wavelength of the light, the optical system will be "sensibly perfect." For a Newtonian telescope, this means that the glass surface must be within one-eighth wavelength of a geometric paraboloid since the light traverses any surface error twice. Since the wavelength of yellow-green light is 22 millionths of an inch, the glass surface must approximate a true paraboloid to within three millionths of an inch or better according to Rayleigh.

But let's examine Rayleigh's criterion critically: In particular, what did he mean by "sensibly perfect"? We can calculate now, as Rayleigh could not, the effect of errors in the optical wavefront, and what we find is that a smooth ¼-wave error reduces the amount of light in the central Airy disk from 84 percent of the total (with the remaining 16 percent in the diffraction rings) to 68 percent, the other 32 percent going into the rings.

Rayleigh specified a ¼-wave because the Airy disks of two close star images that appear just separated with a perfect telescope appear to blur together when viewed in a telescope that just satisfies this criterion. It is, on the whole, a fairly sensible rule and fine as far as it goes. "Sensibly perfect" means "pretty good, but enough less than perfect that a critical observer can see the difference." The light that is not focused into the Airy disk spreads instead into the diffraction rings, reducing image contrast and quite possibly placing fine planetary detail below the contrast threshold of visibility.

Halving the quality criterion to ⅛-wave results in a marked improvement—now 80 percent goes into the disk and 20 percent into the rings. Again halving it to 1⁄16-wave, we finally produce the all-but-perfect diffraction image, with 83 percent in the disk and 17 percent in the rings. For critical applications, though, the errors in each individual mirror or lens in the telescope must be even less than 1⁄16-wave because the errors of the components add up.

However, it's not really possible to use one number to characterize a mirror's surface because errors occur on different scales. The shape of the glass surface *as a whole* is the "figure" which is more or less what Rayleigh was talking about. With small and medium-sized optical systems, we determine the figure by taking a dozen or so samples across the diameter. Fine details go undetected, or at least do not appear in the Foucault graph.

Lumpiness is a medium-scaled error. The surface may have an excellent figure, but performance may suffer from regular or randomly-placed high and low regions ranging from an inch on down. If they are random, they're called "dog biscuit"; if they are regular and concentric, they're called "zones." Such hill-and-valley irregularities don't greatly affect the gross appearance of the Airy disk and inner rings, but they do introduce wavefront errors that throw a significant amount of light into the outer rings and an extended region around a star's or planet's image. Lumpiness obviously degrades planetary detail.

Lumpiness is typical of reflectors but rare in refractors. This is the principal reason that refractors often out-perform reflectors for observations requiring superb resolution. Lumpiness arises from too-rapid polishing, poor contact with the lap, local sticking of the tool, defects in pitch facets, and multiple, poorly blended figuring steps. Polishing slowly with a good lap in good contact generally produces a smooth figure.

Finally, millimeter-scaled roughness, called "micro-ripple," scatters light over many tens of times the diameter of the Airy disk. A very severe case will cause a halo around a bright star, much as dirt on the mirror does. Microripple is caused by small irregularities in the tool, coarse polishing agents, and too-fast polishing. Microripple is seen in commercial optics (such as eyeglasses) polished on cloth polishers, and sometimes on irregularly curved surfaces such as corrector lenses in Schmidt-Cassegrainian telescopes, but it is not often seen in properly polished, handmade, home-brewed optics.

Before concluding that super-smooth, 1⁄16-wave optical surfaces are an absolute necessity, or that good optics alone guarantee good images, consider some other factors that influence image formation. The 30 percent aperture obstruction typical of a Cassegrainian or Schmidt-Cassegrainian telescope produces as much image degradation as a smooth ¼-wave figure error does. Furthermore, temperature variations of a few tenths of a degree in the air inside the telescope tube introduce wavefront errors exceeding ¼-wave. And before starlight reaches the mirror, it must pass through *miles* of inhomogeneous air. So if you wish the ultimate in optical performance, an excellent mirror is not enough; you must keep the aperture obstruction small (which we have done with the telescopes in this book), control the air currents inside the telescope tube (the plywood, spiral-wound paper, and Bakelite tubes are good), and place your telescope at the best possible site.

There are few limits to the pleasure a determined observer with a good telescope can enjoy. Certainly there is a magnitude limit imposed by the collecting area of the mirror, a resolution limit imposed by the aperture, a "seeing" limit imposed by atmospheric turbulence, a "magnificence" limit imposed by light pollution; but there is no limit to the observer's enthusiasm and imagination.

PICKING SITES

Most people don't have a lot of choice in their primary observing site. It must be close to home, accessible, reasonably dark, and fairly secure. This usually means the backyard for homeowners, the rooftop for apartment dwellers, or a nearby park or vacant lot. Count yourself lucky if the sky is dark overhead and there's a grassy area close by without lighting that glares in your eyes.

Open, grassy areas are good because grassy ground heats slowly and cools quickly. Avoid concrete and asphalt—these heat up in the daytime, and warm air rising from them and mixing with cooler air creates turbulence at night. Avoid observing across roofs, over chimneys, vent stacks, and other sources of warm air. Cities often display remarkably good seeing— just the thing for lunar and planetary observing— because the mass of heated air over them becomes trapped in an inversion.

Try to find a reasonably unobstructed horizon. In the East, the South, and the Midwestern United States, trees often obscure part of the sky, in which case there isn't much you can do but look for open spots and clearings. Remember, though, that you can pick up and carry a portable telescope around a backyard or a small clearing to get at different sky areas.

It's nice to observe from a spot that's protected from the wind. On a cold winter's night, wind can make observing almost impossible; even in summer, bring along a jacket. If you become serious about observing, think about constructing an observatory or observation shelter.

Probably nothing detracts so much from the pleasure of astronomical observation as light pollution. If you live in a city, light pollution will limit your

Observing with Your Telescope
12

Photograph by Ronald Royer.

observing to bright sky objects. For faint objects, take evening and overnight trips to parks outside the city, visit friends who live under dark skies, or observe from the grounds of an astronomy-club observatory outside the city. The best time for dark-tripping is close to the time there's a new moon—there is no point in trading city light for moonlight. Light pollution spreads far, so for truly dark skies, your site must be 60 miles from the nearest city of 1,000,000 population, 30 miles from any city of 60,000 or more, and well away from military bases and mining operations, which tend to be brightly lit all night.

Even in the wilds, outdoor lights are a significant annoyance. Otherwise pitch-black mountain and desert parks, where the sky overhead is magnificent, often run dozens of sodium and mercury-vapor lamps all night. Let the rangers know you intend to set up a telescope; they can usually tell you the best places from which to see the sky, and they may allow you to go into unlit areas normally off-limits after dark.

A WORD ABOUT MAGNIFICATION

Which magnification is the most appropriate? It varies, depending on what you're looking at, the telescope's specs, the seeing conditions, and the acuity of your eyes. For sweeping the sky or searching for galaxies and nebulae, use low power. Exit pupils between 5mm and 7mm, or 3.5x to 5x per inch of telescope aperture, give the widest fields, the greatest ease in identifying star fields, and the best views of bright galaxies and clusters.

For general observing, for showing off the moon and planets to friends, for looking for star clusters and observing deep-sky objects in light-polluted skies, the medium-power range between 5x to 15x per inch of aperture gives the best results. The exit pupil is small enough (5mm to 1.5mm) that the image looks sharp and crisp but is big enough that it also appears bright and real.

Depending on your vision, between 15x and 40x per inch of aperture, or between a 1.5mm- and 0.6mm- exit pupil, you'll be able to see all the detail that your telescope resolves. This is the high-power range.

More magnification is seldom necessary or desirable.

Finally, of course, there is the 40x- to 60x-per- inch range, best suited for observing close double stars, observing the brighter planets on nights of exceptional seeing, and testing and aligning the optics of the telescope. Everybody wants to try very high magnification, even if just for the heck of it, so pur- chase a good short-focus eyepiece to try splitting double stars or observing Mars at 500x.

THE OBSERVER'S EYE AND STATE OF MIND

Amateur astronomy is, to a large extent, the art of seeing applied to celestial objects. To get satisfaction from your telescope, you must learn to *observe*, that is, to see with intelligence and understanding, to translate what you see into a permanent and rea- sonably impartial record, and to compare what you see now with what you saw before. There is no formula: You learn it by doing it.

Pamper your eyes and mind physiologically and psychologically. Illness, exhaustion, cold, vitamin deficiency, irritants, alcohol, and emotional strain or agitation reduce the eyes' ability to see and the brain's ability to interpret images. Try to approach an observ- ing session rested, relaxed, and emotionally fresh.

After gathering eyepieces, star charts, flashlight and other gear, and setting the telescope outside to cool, spend five to ten minutes in preparing an agenda for the evening. Work under dim light in a quiet place to start preparing your eyes for the dark and your mind for fresh impressions. Now, what do you intend to look for? Initially, of course, you'll have very broad goals. You might plan to observe the moon and identify several craters, then inspect several bright planets that are up, then search for six or eight bright Messier objects.

Write the agenda in your observing diary, and as you observe, make quick notes on, or sketch, what you see. The desire to describe what you see will encourage you to inspect rather than glance, and describing helps you interpret what you see. Naturally you'll also want to include notes on the weather,

your telescope, the eyepiece used, the seeing conditions, as well as your observations.

Be totally honest. Don't be afraid to write "Unable to find M-31. Where is that &%$#*@ galaxy?" or "Saturn beautiful at 150x. Temperature -10°," accompanied by an ugly sketch. It will be a start, a record

Satellite photographs reveal the extent of outdoor night lighting. USAF photograph.

you may treasure years later—and eventually you will find M-31.

As you become more experienced in observing, set more specific goals, such as sweeping up 20 or 30 new galaxies per night, locating asteroids, or making estimates of variable stars for the AAVSO (American Association of Variable Star Observers). Don't be afraid to try new things.

Remember that your eyes will need to be fully dark-adapted before they are maximally sensitive to faint astronomical objects. Fill the 30 to 45 minutes this takes by observing brighter things, by learning a few more constellations, or by splitting a few double stars just for the fun of it.

OBSERVING THE MOON

No other object in the sky presents such a challenging wealth of detail as the moon. The moon is ½° in angular diameter, or some 1800 arcseconds. Compare that to Jupiter's 45 arcseconds, Saturn's 20, or Mars' 25, for some idea of how much the moon offers.

Equip yourself with a chart of the moon's surface; then follow the lunar terminator as it moves night by night across the roughly cratered highlands and the low, smooth lunar maria. Perusal at low power helps to orient you. First, note several prominent features and learn their names. Then switch to high magnification for a careful inspection. Compare what you see by eye, with binoculars, and with the telescope at various magnifications. On a night of good seeing, the moon is one of the most rewarding objects to observe with high magnification.

As the moon waxes toward full, contrast flattens out. Follow a few features through a lunar day. One of my favorite crater groups is Theophilus, Catherina, and Cyrillus, nestled on the rim of Mare Nectaris. These three, each about 60 miles across, are on the terminator five or six nights after the new moon. On the first night they appear, one of them may still be shadow-filled, but by the next night, the lunar morning is well advanced and shadows are short. Start sharpening your observing eye—take a pad of paper and a soft pencil with you to the eyepiece and sketch what you see. Within a few nights, you will notice that small craters that looked like deep holes will

flatten into nearly invisible rings. As the moon approaches full, these will turn into bright spots surrounded by rays.

It will probably take you the better part of a year to become familiar with the lunar surface and to learn the names of the several hundred craters, the largest maria, mountains, rilles, and rimae. Craters, incidentally, are named after famous scientists, the "seas" after fancied weather conditions (e.g., Sea of Clouds, Sea of Rains, Sea of Tranquillity), and the mountains after those of Earth (thus the Alps, Carpathians, and so on). Learning the names is a good way to learn about the geology and history of the moon, and you'll soon begin to recognize them as familiar friends.

Of course the lunar surface has been pretty well mapped from lunar orbit, so you aren't likely to make any dramatic discoveries while observing it. You can, in fact, buy a photographic atlas of the moon (including the backside) that shows a lot more than any telescope can see from Earth. Use the moon as a source of observing pleasure and intellectual challenge, and to sharpen your observing skills.

Finally, don't forget that the moon probably will impress your non-astronomical friends more than anything else you might try to show them. When they wonder out loud why you built a telescope, promise them the moon—and then deliver!

THE BRIGHT PLANETS

Four of the planets are prominent. They are Jupiter, the giant planet; Saturn, the ringed planet; Mars, the red planet; and Venus, the cloudy planet. You should already know where they are from learning the bright constellations. If you don't, refer to a monthly astronomy magazine.

Jupiter is in many ways the most interesting of the bright planets to observe. It displays the continuously changing order of its four bright satellites (discovered by Galileo almost immediately after the application of telescopes to astronomy) and the rather subtle phenomena of its cloud belts.

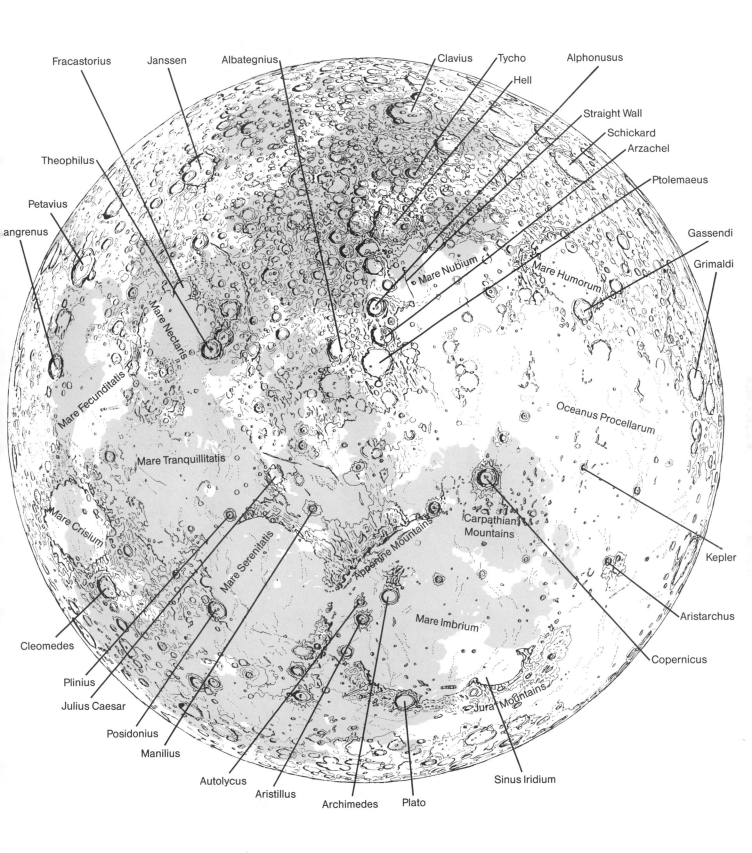

The moon rewards observers with rich and exciting detail.

Jupiter's cloud belts are visible even in small telescopes.

Telescopically, Jupiter is a yellowish or cream-colored oval typically 47 arcseconds equatorially and 44 arcseconds across the polar diameter. Jupiter rotates fast enough that cloud features move on the disk in less than half an hour. These features are pastel in color and low in contrast—don't be fooled by the computer-enhanced images returned by *Voyager!* To an untrained eye, as yours will be at first, the darkest cloud belts will appear as faint gray or brown smudges. The Great Red Spot, a long-lived oval storm in the northern hemisphere of the planet, is about one-fifth the diameter of Jupiter and varies from dull orange to pale reddish-brown in color. Don't expect to see it at first: There's a fifty-fifty chance it won't even be on the side of the planet facing you, and if it's in a pale phase, you probably won't see it until you get your "observing eyes."

The Jovian moons provide a continuously changing show. From inner to outer, their names are Io, Europa, Ganymede, and Callisto. Their orbital periods range from 1.7 days for Io to two weeks for Callisto. The four Galilean satellites are readily visible by all of the telescopes described in this book; the remaining dozen or so are too dim to be seen without a large instrument. If you observe the system for a week or two and record their positions, you can deduce which is which, much as Galileo did when he discovered them.

The best way to become sensitized to planetary detail is by sketching. Put in a medium-powered eyepiece (say, 80x with a 4″, 120x with a 6″, or 160x with a 10″), sit on a box or chair, and look and look and look. Spend half an hour or more just looking. As your eye accommodates to the glare of the planet's disk and the pastel features on it, you'll start to see. Draw what you see, roughly, quickly—visual notes for yourself. Make a dozen sketches. At first, don't draw shadings, just capture the forms. Look for broad, darker bands, lighter belts, dark and light ovals, swirls, and ligatures joining bands and belts. Try shading when you feel you're up for it.

Saturn is the outermost of the planets known since antiquity and is far more beautiful than any other. The planet appears as a flattened, yellowish oval, smaller

(20 arcseconds equatorial by 18 arcseconds polar diameter) and blander than Jupiter but surrounded by a broad, bright ring unparalleled elsewhere in the solar system.

The ring is made of blocks of ice (ranging from a few inches to some tens of feet across) that are gravitationally confined within a few hundred feet of a ring plane hundreds of thousands of miles across. The *Voyager* spacecraft showed that there is an enormous amount of fine structure in the ring, breaking down, as resolution increases, into hundreds and then thousands of narrow ringlets. In a telescope, however, the ring system appears to be smooth and fairly uniform, and divided into three major parts. The outermost A ring is separated from the broader, brighter B ring by a narrow gap called Cassini's Division. Under good conditions, the 4″ reflector shows Cassini's Division quite well. Inside the B ring is the dusky C ring, or crepe ring. The C ring usually yields only to superb optics, good seeing, and well-trained eyesight.

Our view of the rings is not constant. The ring plane is inclined some 27° to the ecliptic, so that as Saturn moves along its orbit, we see one side of the ring for about 15 years, then we see it edge-on, then the other side for another 15 years, and then edge-on again. It was last edge-on in 1980 and will be at maximum presentation of the northern face in 1988. It will be edge-on again in 1995 and reach maximum southern presentation in 2003.

Saturn takes magnification well—better, in fact, than Jupiter. Look for the subtle color-and-brightness changes between the ring and the ball of the planet; these changes depend on the inclination of the ring. Look for the shadow of the ring on the planet and the shadow of the planet on the ring. Can you see Cassini's Division? Search the ring surface for brightness variations. Compare the brightness of the A and B rings. Can you see the crepe ring? Search the ball of the planet for bright and dark bands and cloud features.

Saturn is quite possibly the most beautiful of the planets. Photograph by Jean Dragesco.

Mars reveals its polar cap and subtle dusky surface markings at biennial oppositions. Photograph by E. Ken Owen.

Saturn has a least 16 satellites, but only one bright one, Titan. Titan is eighth magnitude, orbits the planet every 16 days, and can be seen with nearly any telescope. Closer to the planet is Rhea (observable with the 4″); then, in descending order, Tethys, Dione, Iapetus, and Enceladus, all reachable with the 10″ reflector. Although it's impossible to confuse the Galilean satellites of Jupiter with anything else, Saturn's satellites can be confused with background stars. To figure out which is which, refer to the *Astronomical Almanac*.

Mars comes to opposition (that is, opposite the sun in the sky) once every 26 months. Because its orbit is eccentric, some oppositions are better than others:

At the best oppositions, or once every 17 years, its telescopic disk swells to 24 arcseconds; at a poor opposition, to roughly 14 arcseconds. (The next close opposition will be in September 1988, the next poor one in February 1995.) More than three months before or after opposition, the visible disk is too small to show much detail; indeed, even at opposition, Mars is a rather difficult object telescopically.

Don't let this prospectus discourage you: When Mars next moves toward opposition, observe it. The "Red Planet" appears as a soft orangey-pink with pale gray-green darker areas and yellowish lighter regions. The poles may be distinguished by markedly brighter and whiter polar caps; they are the easiest feature, by far, to see. Although Mars has a hard surface, its markings are not constant. Winds blow fine surface dust into the atmosphere, clouds form and dissipate, and the polar caps partially evaporate and re-form each martian year. You can spend many enjoyable evenings sketching Mars when it is near opposition, then constructing a map of the planet based on your sketches.

The 10″ reflector or 6″ refractor qualify as instruments suitable for serious Mars observing. You will certainly see, without trouble, the large, dark markings such as Syrtis Major and Sabaeus Sinus. If a big dust storm blows up, as one did for the approach of NASA's *Mariner 9* orbiter, you can spot it by its yellow tinge and night-to-night spread over the planet.

Venus, shining in the evening sky at greatest western elongation, beckons brilliantly. The telescope, however, reveals little about Venus but its featureless cloud deck and changing phases. Because its orbit lies inside Earth's orbit, Venus never appears more than 48° from the sun and scoots through inferior conjunction very quickly. From greatest western elongation, it sinks rapidly in the sky, becoming a slender crescent considerably larger than Jupiter. Venus finally disappears into the sunset, only to reemerge in the dawn sky and move toward greatest eastern elongation 140 days later. Growing smaller and fuller, it passes through superior conjunction, reappearing in the evening sky, and grows thinner and larger, to reach greatest western elongation again 584 days after starting the cycle.

Venus is bright enough to find in the daytime sky if you use a mounting with setting circles. Daytime observing, which suppresses the overwhelming glare of the planet, sometimes permits seeing faint detail, usually streaks or ill-defined dusky patches, on the cloud deck. At inferior conjunction, Venus shines as a ring of light—the thin crescent completed by light shining through the atmosphere of the dark side.

THE ELUSIVE PLANETS

Any attentive nature-lover will discover the four bright planets in the course of casually observing the sky. To find the remaining four planets, you need advance knowledge. Mercury flits back and forth from the

morning to the evening sky, appearing as a faint morning or evening "star." Uranus is marginally visible to the naked eye; it was the first planet discovered with a telescope. Neptune is well below naked-eye visibility, and you will need a good chart to find it. Pluto is nearly lost among myriads of faint stars; it is only just visible with the 10″ reflector.

Mercury swings around the sun every 88 days. Like Venus, it is always near the sun—never more than 28° away. Mercury, therefore, is seen only shortly before sunrise or shortly after sunset. Because Mercury is a small planet and has a rather dark surface, it is not

Venus's principal attraction lies in its phases—an early proof of the Copernican solar system. Photograph by Jean Dragesco.

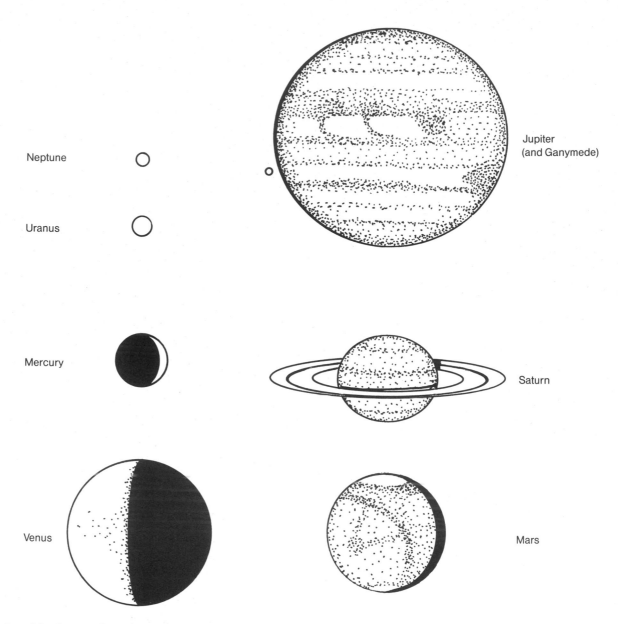

Relative sizes of the planets under optimum viewing conditions.

particularly conspicuous even when it is well-placed for observing. Small wonder that many professional astronomers have never seen it.

Our view of Mercury suffers from having to observe the planet through a long, turbulent air path. Its disk, at times of greatest elongation, is 8 arcseconds across.

Consider yourself lucky if you spot the phases through low-altitude air. Like Venus, Mercury can be observed during the daytime, but because of its dark surface, it appears a washed-out yellow-brown against the blue of the sky.

Uranus was first recognized as something other

than a star by William Herschel, the great English amateur-turned-professional astronomer, in 1781. He was examining stars with a 6″ reflector he had made himself. Uranus looked unusual; he could see its disk. Herschel announced his discovery as a comet, but its orbit soon showed the object to be another planet.

Uranus appears nonstellar at 60x to 80x. It is magnitude 5.5, greenish to a practiced eye, and has a disk of 3.5 arcseconds across, about one-twelfth the angular size of Jupiter. To find it, use the finder chart from a monthly astronomy magazine or Ottewell's *Astronomical Calendar*. Locate the field in your finder, then examine the stars. Uranus is the greenish one with the tiny disk. If this approach doesn't work, step up the magnification to 100x to 200x. Telescopically, Uranus does not offer a great deal. The 4″ and 6″ reflectors show that the planet has a disk. The 10″ shows the disk clearly, although without detail, and will reveal Uranus's brightest satellites to the determined observer.

Neptune is one and a half times farther from the sun and two magnitudes dimmer than Uranus. This eighth-magnitude planet appears only as a tiny, blue-green disk. To find it, you'll need a chart for the current year, and be prepared to spend some time in star-hopping to it. Once you've found it, switch to higher magnification to see the planet's disk. Neptune is recognizable with the 4″ or 6″ telescopes, but switch back and forth between the planet and nearby stars to convince yourself that the image is nonstellar. Its disk and color are readily apparent with the 10″.

Pluto moves around the sun in an eccentric orbit that will bring it nearest the sun in 1989, and nearer the sun than Neptune between 1979 and 1999. Pluto is much smaller than Neptune and therefore much dimmer. It shines at roughly 14th magnitude, and its disk cannot be distinguished by any Earth-bound telescope; it remained lost among myriads of faint stars until 1930.

To find Pluto, plan on a careful search with a good chart and with the 10″ telescope. Don't consider it found until you've observed it moving from night to night against the background of stars. Experienced observers claim to have seen it with 6″ and 8″ telescopes, but only after having found it in larger in-

struments. Pluto looks unexciting—a mere speck of light moving among the stars—but repays the effort of the search with a deep and lasting thrill.

ASTEROIDS

Between the orbits of Mars and Jupiter lies a void empty of major planets. It is, instead, filled with several thousand small bodies called asteroids. The four largest and brightest—Ceres, Pallas, Vesta, and Juno—reach opposition every 16 months or so and come close to naked-eye brightness. None displays a telescopic disk; as the name implies, they appear "starlike." Finding them is not at all difficult; plot the path on a star atlas, then go star-hopping to the charted location. Sketch the field of stars as carefully and accurately as you can. Several nights later, find the "star" that moved. That one is the asteroid.

Coordinates for the "Big Four" appear in monthly astronomy magazines and the *Astronomical Almanac*; ephemerides (lists of coordinates by date) for fainter asteroids are published in *Tonight's Asteroids*. Asteroids are fun to observe, enough so that some observers have located hundreds of them. While you may never become that thrilled with them, it's nice to add them to the list of solar-system bodies you've seen.

COMETS

Once every five or ten years, a bright comet swings around the sun, stretching like a ghostly apparition across the sky. Bright comets are impressive. Their nucleus usually appears starlike, but you may see jets of material spraying into the coma, changing hourly, with knots and kinks moving night by night down the tail. The tail may stretch 10° or more across the sky and be viewed with the naked eye or through an ordinary 35mm camera.

The biggest comets tend to be the "fresh" ones that have never encountered the sun before, but their appearance is intrinsically unpredictable. About half of all such comets are discovered by amateurs who are searching for them, usually with telescopes of between 4″ and 8″ aperture. When you hear of the

Comet West appeared unexpectedly from the outer reaches
of the solar system in 1976 and was visible for several months.

discovery of a bright comet, drop everything else to observe it. It may be bright and visible for a few weeks before its orbit carries it away from the sun and it fades.

STARS

Stars are something of a disappointment. In a good telescope, stars appear as tiny, hard points of light. It's easy to conclude that when you've seen one star, you've seen them all. But this doesn't need to be the case. The key to observing stars is not to look for detail, but to concentrate on their light. Astronomers use photometers and spectroscopes to measure the intensity and spectral composition of starlight. While you, as a neophyte observer, cannot do this, you *can* easily observe the principal variation between stars: their colors. Stars come in a wide diversity of colors—from the yellow-orange of the cool type-M (surface temperature, 3000°K) through a range of subtle yellow tinges to white (at roughly 10,000°K) and on to the blue-white sparkle of the B and O supergiants (temperature, 30,000°K plus).

Because our eyes are not sensitive to color at low light levels, star colors are unfamiliar to most non-astronomers. Your telescope, in gathering hundreds of times the light that unaided eyes can, brings bright stars into the range where you can readily detect the colors. Start by observing bright red and blue stars—Betelgeuse and Rigel, for example—moving from one to the other and comparing their colors. Later you may want to search out some of the stars with more extreme colors. Mu Cephei, also known as Herschel's Garnet Star, is one of the reddest stars in the sky. Its blood-red color is astonishing in any telescope. Compare it with nearby Beta Cephei or Mu Columbae, the latter one of the bluest stars known. *Burnham's Celestial Handbook* contains a wealth of information about individual stars.

Double stars and other multiple-star systems contain between half and one-third of all the stars in space. These stars are actually in orbit about one another. Some of the closer pairs have made several complete orbits since their discovery. In perusing lists of stars, you'll find many stars noted as double, but not all of these can be seen as two with a tele-

scope; many are too close together and known only from mutual eclipses, mixed-type spectra, or radial velocity variations.

Doubles frequently consist of stars with strongly contrasting color, and they can be used as a test of the telescope's optics and the seeing quality of the night. Observers' handbooks are filled with lists of doubles. You will soon become acquainted with the easy gold-and-yellow pair, Albireo, or Beta Cygni; the famous "double-double," Epsilon Lyrae, with its close 2-arcsecond and 3-arcsecond pairs separated by some 200 arcseconds; Mizar, the second star from the end of the Dipper's handle, divided by some 14 arcseconds; the quadruple Iota Orionis at the heart of the Orion Nebula; or the difficult companion of Rigel, some 10 arcseconds away but 8 magnitudes dimmer than its primary.

Variables are stars that change brightness. They may change for a variety of reasons: eclipses between components of a binary pair, pulsations of an unstable star, flares and other eruptions, star spots on their rotating stellar disk. The reason for observing them is that many variables are not yet understood; before they are, their variations must first be characterized. Normally you can't look at a variable star and expect to see any change right away. Instead, measure or estimate its brightness, record that quantity, then measure or estimate it again several days or weeks later.

To estimate the brightness of a variable, compare the variable with other stars in the same field of view. Typically, the observer assigns them arbitrary scale values; then by looking from comparison star to variable to another comparison star, he estimates what the scale value of the variable should be. This process is fairly reliable and consistent—it gives accuracy on the order of 10 to 20 percent—and, best of all, it requires no extra equipment. Observers with small telescopes equipped with home-built photometers can measure the brightness of a star to better than 1 percent but cannot produce as many measurements as "eyeball" observers can. The American Association of Variable Star Observers (AAVSO) distributes charts for stars, as well as collecting and collating millions of estimates that it makes available to astronomers. Membership is open to amateurs.

You can sample variable-star observation without a

telescope. Make nightly estimates of Delta Cephei's brightness relative to nearby Epsilon and Zeta. Designate Zeta's brightness as "2" and Epsilon's as "8"; assign the variable a value on a scale of 1 to 10. (The whole process, including writing the estimate in your observing notebook, should take about 30 seconds per night.) After several months of making estimates, plot them against time and then see if you can deduce the period of the star's variation.

THE MILKY WAY

Every telescopic object (except other galaxies) is part of the Milky Way, which is the underlying structure of the visible heavens. The center of the galaxy lies in the direction of Sagittarius. The Milky Way is at its broadest and brightest in this area, and telescopically the area abounds in star clusters.

To the naked eye, the galaxy is a softly glowing band of light that circles the entire sky, rising in Scorpius and Sagittarius, flowing through Aquila, Cygnus, Cepheus, Cassiopeia, Perseus, Auriga, past Orion, then down into Canis Major, Puppis, Vela, Crux, Centaurus, Norma, and around to Scorpius and Sagittarius again. Observe it carefully and you'll see considerable structural detail: the large star clouds of Sagittarius and Cygnus, the small star cloud in Scutum, the Great Rift (which stretches from Sagittarius to Cygnus), and the Coalsack (in the southern hemisphere). Sweep along the Milky Way with binoculars or your telescope at low magnification to find dozens of small clusters of stars, nebulae, and the most star-filled fields in the sky.

Our sun lies in the galactic suburbs, along the inner edge of a ragged spiral arm. Sagittarius marks the direction to the galactic center, galactic longitude 0°. Perseus and Auriga, relatively star-poor regions, lie in the anti-center direction, or 180°; the sparkling star-fields of Cygnus lie at 90°; and the riches of Vela lie at 270°. Few external galaxies can be seen near the galactic plane; their light is blocked by dust and gas.

Our galaxy is a fine sight by naked eye and a rich ground for telescopic observation.

To see other galaxies, look away from the Milky Way and into the star-poor regions of Leo, Virgo, Ursa Major, Cetus, Aquarius, and Sculptor.

As you use various lists, charts, and handbooks in your observing activities, note where objects lie relative to the Milky Way and you will begin to understand that most telescopic objects are the component parts of a grand scheme visible without telescopic aid.

STAR CLUSTERS

Sprinkled throughout the Milky Way are numerous clusters of stars. You can find them in star-rich areas— Sagittarius, Scutum, Auriga, Gemini, Canis Major, and Puppis—simply by sweeping at low magnification. The brightest clusters appear as a field full of brilliant stars; the dimmest, as a soft glow with a few barely discernible pinpricks superimposed.

There are actually two types of clusters, each quite different in distribution and appearance. *Galactic* clusters (also known as *open* clusters) consist of a few dozen to a few thousand loosely associated stars; they are embedded in the spiral arms of the galaxy and appear in or near the Milky Way. Some, such as the Pleiades and Hyades, can be seen with the naked eye. Others, the Beehive cluster in Cancer or M-44 in Canis Major, can be seen with binoculars as little glows. Even a small telescope will show dozens of open clusters strung along the Milky Way.

The other type of star cluster is the *globular* cluster, a much more massive cluster, consisting of hundreds of thousands of stars in a relatively compact swarm or ball of stars. They concentrate strongly in the direction of the galactic center; virtually all of them are within 90° of it. Sagittarius, Scorpius, Ophiuchus, Norma, Ara, Telescopium, Corona Australis, and Microscopium—the constellations nearest the galactic center—are loaded with globular clusters.

With the 4″ telescope, globulars barely reveal their stellar nature. Under good conditions, sharp-eyed observers can just discern stars against the background glow of unseen stars. A 6″ telescope "resolves" the brighter globulars; with the 10″ telescope, a bright globular (such as M-13 or M-22) is a starry ball.

The double cluster in Perseus is a magnificent sight with the 4″ reflector.

NEBULAE

Stars form from clouds of dust and gas; they throw off shells of gas as they age, and some of them die in colossal explosions that throw star-stuff across space. All these gassy, dusty clots of matter are called nebulae. The word "nebula" means cloud in Latin, and indeed, some are rather cloudlike in appearance, but the term originally included many objects that today we know are stellar, such as galaxies. We now distinguish five types of nebula: diffuse, reflection, dark, planetary, and supernova remnants.

Diffuse nebulae are the glowing, red clouds that often grace textbooks and telescope sales brochures. They are usually regions of star formation, and they glow by absorbing and reemitting light given off by newly formed stars. Telescopically, little or no color can be seen—our eyes are simply not sensitive to

In the 10″ reflector, a globular cluster such as M-22 appears as a ball of stars. Photograph by Thomas Dessert.

Nestled in the heart of the Milky Way, the Lagoon Nebula, M-8, is the finest diffuse nebula in the summer sky. Photograph by Thomas Dessert.

color at such low light levels. Nebulae usually appear as a pale, almost ghostly, gray color, perhaps greenish if you're straining to catch color. They concentrate in the plane of the Milky Way and often have star clusters embedded in them. Nebulae are, in fact, where star clusters come from. The brightest are the Orion Nebula, M-42, in the sword of Orion; the Lagoon Nebula, M-8, in Sagittarius; and the Eta

Carinae Nebula, NGC-3372, an object for Southern Hemisphere observers.

Reflection nebulae, on the whole, are far less well-known than diffuse nebulae. They are clouds of "dust" —finely divided solid material between the stars, reflecting light from those stars. The faint nebulosity in the Pleiades star cluster is a reflection nebula.

Dark nebulae were discovered early in the twentieth

M-57 is the famous "smoke-ring-in-the-sky" ring nebula in Lyra; a small but relatively bright deep-sky object. Photograph by Martin Germano.

century but acknowledged only some 30 years later. Dark nebulae are diffuse or reflection nebulae that are not lit up. They abound in the plane of the Milky Way but are seldom visually observed. A determined observer, however, can find several dozen from a dark-sky location with a 10″ reflector.

On a grand scale, the Great Rift in the Milky Way is a huge network of dark nebulae. The "fishmouth" feature in the Orion Nebula is a dark nebula, and the famous Horsehead is one silhouetted against a diffuse nebula.

Planetary nebula have nothing to do with planets; rather, they are short-lived shells of gas thrown off by aging stars. They are round, or at least display some symmetry, and have rather dim disks from several arcminutes down to a few arcseconds. The best-known planetary is the Ring Nebula, M-57, in Lyra. In a telescope, it looks like a faint oval glow about twice the apparent size of Jupiter; telescopes larger than 4″ show it to be dimmer in the center. Showpiece planetaries include the Dumbbell, M-27, which is large and bright; and the Owl, M-97, which is faint.

Supernova remnants include the Crab Nebula, M-1, a violently twisted object once thought to be a planetary; and the Cygnus Loop, an open ring of diffuse nebulosity that is the expanding shell left by a supernova explosion. The Crab is just a faint patch in the 4″ and 6″ telescopes, but with the 10″, it begins to show some of the famous filamentary structure seen in photographs. The Cygnus Loop is a difficult object. 52 Cygni lies close to one side of the Loop. From a dark-sky site, the nebula is a faint, glowing filament in the same field.

GALAXIES

Nothing you will observe looks so faint, yet retains its ability to impress, as a galaxy. Galaxies are distant systems of billions of stars made barely visible by distances measured in millions of light-years. If looking at galaxies doesn't make you feel small, nothing can.

When observing with the largest telescopes, there appear to be as many galaxies as there are stars. For telescopes in the 4″ range, there are only roughly a dozen galaxies worth looking at, that is, offering

more to see than a faint glow. The number of visible galaxies grows rapidly with increasing aperture; there are several hundred interesting prospects for the 10″ telescope, and several thousand for a 16.″

Begin by observing the most famous galaxies—the Andromeda galaxy, M-31, a first cousin of our own system and looking like a faint oval cloud; NGC-253, a southern spiral; the Magellanic Clouds (which can't be seen from the U.S.); and M-101 and M-51, which show hints of spiral structure in 6″ to 10″ telescopes.

Galaxy-observing is an activity for connoisseurs. Galaxies are faint objects easily overcome by minute amounts of light pollution. They require big apertures and well-developed observing skills in order to be appreciated, but given proper attention, they reward the observer well. Details such as the globular clusters in M-31 and M-33 can be ferreted out, the spiral arms and dust lanes traced, and their starclouds and nuclei inspected.

ASTROPHOTOGRAPHY

Astrophotography became a major amateur activity once fast films and fast lenses were available. It is a simple matter to load a 35mm camera with fast film, set it on a tripod, open the diaphragm wide, focus to infinity, set the shutter to time, and record more stars in five- to ten-second exposures than you can see by eye. Images of the Milky Way require exposures of ten minutes at f/2.8 on today's fast films, too long for stationary camera exposures. You can piggyback the camera on an equatorial telescope and track by hand, keeping a star neatly on the cross hairs during the exposure.

Filming the moon and planets is also quite easy. You can even get pretty good results without a clock drive if the exposures are shorter than a quarter-second. The moon is best for initial experiments because it's bright, big, and loaded with detail. For the best results, you'll need a sturdy equatorial mount with a clock drive, auxiliary tubes to increase the effective focal length of the system to 60x or 100x the aperture, and careful technique to avoid shaking the telescope during exposures that run for several seconds.

Your telescope can also serve as a powerful deep-sky camera. Its optical credentials are impressive: A

Faint pinwheels of stars and
galaxies show as dim patches
in small telescopes; film records
more detail in long exposures.
Photograph by Jack Newton.

10″ reflector, rated as telephoto lenses are, is a 1500mm
f/6 ultra-achromat, a stop or more faster than comparable photo gear costing thousands of dollars. In
order to photograph nebulae, galaxies, and other
deep-sky objects, it is essential to keep the telescope
precisely pointed at the object during an exposure
that may run from 20 to 60 minutes. For this, you
need a mount considerably better than those usually
supplied with commercial telescopes. It helps greatly

to do your own photo darkroom work since special
processing, printing, and materials are needed to bring
out details the corner drugstore will miss.

DOING SERIOUS SCIENCE

You may wish to remain a tourist in the realm of
astronomy, happy to see the sights and to point them

The astrophotographer at work: recording the progress of a lunar eclipse.

out to friends and family, or you may wish to become a member of the astronomical community. There are too many stars and too few astronomers, so there will always be a place for observers who wish to contribute to astronomy even though they may be earning their living in another field.

The problem remains, nonetheless, to discover where and how you feel you can do useful and interesting work. The first step is to begin reading the astronomical literature, not just the monthlies, but the publications of specialty organizations. It's also essential to meet other enthusiasts, attend conventions, and talk to people already actively engaged in serious amateur astronomy. Ask yourself: "What are these people doing?"; "What's involved?"; "Can I commit the time to it?"; "Am I interested in this type of observing?"; and "What is the scope for original contributions in this field?"

Most observers settle in one of three areas: 1) estimating or measuring variable star magnitudes, either with the naked eye or with a photometer, in coordination with the AAVSO, IAPPP, or BAA Variable Star Section; 2) timing grazing occultations of stars by the moon or asteroids in conjunction with IOTA; and 3) drawing and photographing planetary features as a member of ALPO or the BAA Planetary Section (see Appendix C: Astronomical Organizations for Amateur Astronomers).

These programs are, for the most part, repetitive observational programs well-suited to the long-term, low-cost nature of astronomy-as-a-hobby observing. The field is rapidly changing, however, and new possibilities are opening up, with the richest and most interesting opportunities probably lying in photometry. The IAPPP, for instance, mounted a special Zeta Aurigae Campaign to monitor the eclipse of that particular star. More programs lie in the offing. Other opportunities are surprising: Professional astronomers need help in ascertaining the rotation rates of asteroids, obtained by monitoring their light curves for periodic variations over several months, which permits deriving the orientation of the asteroid's rotation in space. There are programs to search for galactic novae, novae in globular clusters, and supernovae in other galaxies by eye, photographically, and with television-type sensors. The International Halley Watch is looking for observations from amateur astronomers, in part because 1985 amateur equipment matches professional equipment available in 1910, the last year Halley's Comet appeared.

No one who has built a telescope should hesitate to join these organizations, to read their journals and publications to see if they are interesting, or to strike out on his or her own in search of new ideas. Nor should any individual feel reluctant to maintain an interest in astronomy solely for the pleasure of observing or for the sheer fun of it. After all, astronomy is for *everyone!*

Appendix A
Telescope Materials

PLASTIC LAMINATE

Plastic laminate (Formica), made by gluing together layers of brown paper with a heat-setting phenolic resin, is a perfect material for Dobsonian telescope bearings. It is smooth, hard, and cleanable. Home-improvement centers sell plastic laminate in two forms: as sheet 3/64″ thick, or as finished counter top bonded to particle board 3/4″ thick. Don't buy a full sheet (about $40) for your telescope; purchase leftovers from kitchen counter tops—"sink cutouts"—for just a few dollars. Get the smooth or slightly rough-surface laminate; avoid the embossed "tile" and "natural slate" textures; they are unsuitable for a telescope bearing. If you cannot locate a place that sells sink cutouts, bond laminate to plywood with contact cement.

TEFLON

Teflon is Dupont's trade name for polytetrafluoroethane (PTFE) plastic. This chemically inert, slick material slides with buttery smoothness against many materials; it is the key to smooth motion in telescope bearings.

To buy it, try the Yellow Pages under "Plastics—Sheet, Tube, and Rod" for a local supplier or one in a nearby city. Teflon comes in sheets and sells for $12 per pound, or roughly $18 per square foot in 1/8″ thick sheets. Since you could build every telescope in this book with that much Teflon, you'll probably get all you'll ever need for less than $10. Ask for scraps or "cutoffs." These usually sell for half-price.

TELESCOPE TUBES

Concrete-form tube is a common and inexpensive material for telescope tubes. Concrete-tube forms are used to cast round concrete posts and pillars. They are sold waterproofed with asphalt and coated with wax inside and outside to prevent sticking. If you can't find them at your local home-improvement center, try big lumberyards or firms that sell concrete. The most common brand name in the U.S. is "Sonotube," but names and properties vary locally. Form tube usually comes in 2″ increments from 6″ up to at least 24″, and usually has a 1/4″-thick wall. It is much cheaper than other telescope tube materials, but it poses several problems. It is rather "squishy," thus weaker than an ideal telescope tube would be, but you can reinforce the ends.

The first step in preparing it is to remove the waxy coating. Do this by using paint thinner and lots of rags or paper towels, by peeling off the outermost layer of paper, or by simply leaving the tube outdoors in summer sunshine for a few weeks until the wax melts and evaporates.

Cut concrete-form tube with a sharp hardware knife (use a sawing motion), a fine-toothed handsaw, or a saber saw with a fine blade. Work slowly to avoid crushing the tube at the place you're cutting. Sonotube is paper, so it must be protected by careful sealing and painting.

Spiral-wound paper tube is brown kraft paper rolled and glued into a rigid tube. For all its apparent cheapness, spiral-wound tube is surprisingly solid and hard to cut. The best tools for the job are a fine-toothed handsaw or a saber saw with a metal-cutting blade. Spiral-wound tube *must* be sealed and painted thoroughly if you live in a damp climate. Use two coats of sealer applied liberally, sanding between the coats. Then apply two coats of a waterproof paint such as polyurethane.

Phenolic tube is made of cloth bonded by phenolic plastic; the resulting tube is light, waterproof, strong, and easy to work. Because phenolic tube has low thermal capacity, it cools quickly and causes little disturbance of the air in the tube. Phenolic tends to be brittle, so cut it slowly and with a fine-toothed saw blade. Sealing is not necessary to keep out water, but since sealers give the paint a better bond, phenolic tube should be sealed before painting.

Fiberglass is a composite material made of fine glass fibers and epoxy-plastic resin. It is light, strong, and durable—nearly ideal for telescope tubes. You can also cast resin and fiber into a mold to make a mirror cell. The disadvantages of fiberglass are that it is messy, smelly, and fairly expensive. The biggest drawback of fiberglass tubes is that unless they are fairly thick, they are relatively flexible (as is concrete-form tube), so a tube made from fiberglass should be reinforced at the ends.

Spiral-wound paper, phenolic, and fiberglass tubes are sold as telescope parts.

PLYWOOD

Buy only exterior-grade plywood for telescopes; interior-grade plywood comes unglued if it becomes wet repeatedly. Plywood faces are graded A, B, C, and D in descending order of quality. Grade A is suitable for staining and varnishing surfaces that will be "on show." For the pragmatic telescope, grade B is entirely adequate. Grade C and D specifications allow open holes in the

veneer—they are suitable only for hidden construction. Interior plies are usually C for exterior plywood and D for interior plywood, although you can buy "marine-grade" (i.e., watertight) plywood with interior-grade B veneer. Each sheet of plywood is specified by two letters—you'll generally want A-A (good on two sides, or G2S) or A-B, although a selected sheet of A-C (good on one side, or G1S) can be satisfactory for a telescope tube where the C side faces in.

WOOD-TO-WOOD JOINTS

Fasteners provide strength to a joint; glue prevents motion that would eventually pull the fasteners out. For telescopes, avoid using nails unless temporarily on wood parts; they pull out. Use wood screws, bolts, or the special fasteners listed below.

Use waterproof glue if at all possible. Yellow carpenter's glue, or aliphatic resin glue, is not really waterproof, but it is convenient and water-resistant. It can be used in dry climates or for parts that will be well-sealed. Casein glues are not waterproof but are sufficiently water-resistant to be used for outdoor jobs. Plastic resin, or urea-formaldehyde glues, are mixed from a powder and are highly water-resistant. When using them, parts must fit together tightly because resin glues do not fill gaps well. Best of all are resorcinol glues, which are completely waterproof and very strong. These, however, must be mixed from liquid and powder components, used within a few hours after being mixed, and they must dry under pressure for about 24 hours.

SPECIAL FASTENERS

The telescopes described in this book rely to some degree on special fasteners. They are available in well-stocked hardware stores, home-improvement centers, and by mail order.

Tee nut: a metal sleeve with a flat collar and sharp prongs that is driven into wood. Tee nuts are a quick and simple way of putting threads into wood. The broad metal collar acts as a washer and a nut, and the prongs keep it from rotating.

Wing nut: a threaded nut with "wings" for turning with fingers. Wing nuts may be used whenever parts must be disassembled with tools or must be readily adjustable.

Threaded insert: a metal cylinder with machine-screw threads inside and special knife-edge threads on the outside for screwing into wood. These inserts are especially useful for installing metal threads in "blind" holes, and they are usually available only at large hardware stores and by mail.

Right-angle connector: a short metal bar centrally threaded through a diameter. These connectors provide the best way to join plywood panels at right angles with high joint strength.

METAL-TO-WOOD JOINTS

Where the wooden and metal parts of a telescope meet, there is the possibility of looseness and motion. Wood dries and shrinks as it ages; metal does not. Wood expands with humidity; metal does not. Wood fibers can crush; metal is rigid. To connect wood and metal, always *clamp wood between metal.* Use bolts rather than wood screws to hold metal against wood; washers or tee nuts on the other side capture the wood between the metal. If the wood shrinks, you need only tighten the bolts and your telescope will be steady again.

FINISHING

A nice finish will not make your telescope perform any better, but the effort is worth the time in appearance. After you have straightened, smoothed, and sanded each piece, you'll see voids and gaps on the edges of the plywood, poorly fixed holes in the veneer, and knots, dents, and nail holes in the surface. Begin by filling large voids, especially gaps in veneers, with a thick mixture of glue and sawdust. Cram the mixture in as far as you can; then let it dry and sand it smooth. Fill small holes with cellulose wood filler. Spread it with a putty knife, working it into the holes and low spots. Sand the surface down to the wood. If there are any remaining holes, fill them, let them dry, then sand again. Wear a filter mask—sanding dust from filler is terrible.

Because telescopes get wet from dew (or drenched from unexpected rain), seal all wood parts, particularly those that hold optical components. Coat the parts, especially plywood edges, with penetrating wood-sealer/primer, sanding between coat and allowing each coat to dry thoroughly. Repeated sandings assure a smooth finish later on. Finally, apply two or three coats of a tough finish color. Virtually nothing is better than polyurethane floor enamel. It is formulated to be scuff-resistant and waterproof. Apply the first coat, let it dry, sand with fine sandpaper, remove the dust with a tack cloth, and then apply the second coat. It may be difficult for you to wait, but allow several days for the paint to harden before you start using the part.

RECOMMENDED ADHESIVES

Silicone rubber adhesive: Dow Corning; sold widely in hardware stores throughout the United States.

Carpenter's wood glue: Elmer's #E-702; "a fast-grabbing, fast-setting, super-strength glue formulated for cabinets, furniture, and other projects." Water resistant, but not waterproof.

Weldwood contact cement: Roberts Consolidated #0515; a nonflammable contact adhesive for bonding plastic laminates (i.e., Formica) to wood.

Pliobond industrial adhesive: Goodyear Tire and Rubber Co.; "flexible, permanent adhesive for wood, fabric, paper, metal, ceramics, glass, plaster, stone."

RECOMMENDED FILLERS AND SEALERS

Carpenter's wood filler: Elmer's #E832; a nontoxic, nonflammable, water-resistant, latex-based wood filler.

Primer/sealer: X-I-M Flashbond #400 Clear; "primes and seals all surfaces" and "makes paint adhere—last longer." An extremely good prime coat but must be used with adequate ventilation.

Gym finish penetrating floor seal: Coast to Coast Hardware #563-4068, 3900-04; "a penetrating, clear sealer designed to protect against warping and splintering of wood. Case-hardens wood to make surface exceptionally resistant to wear and marring."

Appendix B Materials, Suppliers, and Sources

ASTRONOMICAL BOOKS

ASTROMEDIA CORP. (P.O. Box 92788, 625 East St. Paul Ave., Milwaukee, WI 53202). Stocks popular and best-selling astronomical books and star atlases. Source for back issues of *Telescope Making* and *ASTRONOMY*.

ASTRONOMICAL SOCIETY OF THE PACIFIC (1290 24th Ave., San Francisco, CA 94122). Books at discount prices for members of the ASP.

EVERYTHING IN THE UNIVERSE (429 43rd St., Oakland, CA 94609). Select line of books, posters, and astro-products.

HERBERT A. LUFT (P.O. Box 91, Oakland Gardens, NY 11364). Wide range of astronomical and scientific books. Free catalog.

SKY PUBLISHING CORP. (49 Bay State Rd., Cambridge, MA 02238). Carries a line of technical and popular astronomical books and atlases.

WILLMANN-BELL, INC. (P.O. Box 3125, Richmond, VA 23235; Tel: 804-320-7016). Carries "Literature, Electronics, Supplies for the Amateur to Professional Astronomer and Optician," the very best source for both popular and hard-to-find technical astronomy books and star atlases. Prompt service.

ASTRONOMICAL PHOTOGRAPHS AND POSTERS

ASTROMEDIA CORP. (P.O. Box 92788, 625 East St. Paul Ave., Milwaukee, WI 53202). Posters, large illustrated annual calendar, slide sets of planets and stars, astronomical art.

ASTRONOMICAL SOCIETY OF THE PACIFIC (1290 24th Ave., San Francisco, CA 94122). Bumper stickers ("Astronomers Do It in the Dark"), books, posters, slide sets, information packets.

HANSEN PLANETARIUM (15 South State St., Salt Lake City, UT 84111). Wide range of large color nebula and planet posters and slides from professional observatories. Catalog.

LICK OBSERVATORY (University of California, Santa Cruz, CA 95064). 8"x10" and 14"x17" photo prints and 35mm slides. Catalog.

MIRROR KITS

COULTER OPTICAL (P.O. Box K, Idyllwild, CA 92349). Large-diameter thin-mirror blanks from Pyrex.

A. JAEGERS (6915 Merrick Rd., Lynbrook, NY 11563). Basic kit includes blank, plate-glass tool, diagonal mirror, abrasives, pitch.

OPTICA B/C (4100 MacArthur Blvd., Oakland, CA 94619). Catalog "with over 400 items."

N. REMER OPTICS (P.O. Box 306 R, Southampton, PA 18966). Blanks up to 16″ diameter; complete kits up to 12″.

TELESCOPICS (P.O. Box 98, La Cañada, CA 91011). Kits include blank, ceramic tool, six abrasives, pitch, Barnesite polishing agent, and beeswax.

WILLMANN-BELL, INC. (P.O. Box 3125, Richmond, VA 23235; Tel: 804-320-7016). Complete line of mirror blanks, kits, abrasives, and pitch selected for quality.

READY-MADE TELESCOPE PARTS (MIRROR CELLS, ETC.)

KENNETH NOVAK (P.O. Box 69, Ladysmith, WI 54848). High-quality diagonal holders, mirror cells, focusers, ring mounts, etc. Recommended as the best available; very reasonable prices.

MEADE INSTRUMENTS CORP. (1675 Toronto Way, Costa Mesa, CA 92626). Focusers, mirror cells, guide and finder scopes.

NORTH STAR TELESCOPE COMPANY (3542 Elm St., Toledo, OH 43608). Mirror cells, ring mounts, tubes.

PARKS TELESCOPE COMPANY (679 Easy St., Suite B, Simi Valley, CA 93065). Complete line of Newtonian parts, finders, eyepieces, and binoculars. Emphasis on quality fiberglass tubes.

TELESCOPICS (P.O. Box 98, La Cañada, CA 91011). Solid spiders, mirror cells, focusers.

ROGER TUTHILL (P.O. Box 1086, Mountainside, NJ 07092). Finders, polar alignment jigs.

UNIVERSITY OPTICS (P.O. Box 1205, 2122 East Delhi Rd., Ann Arbor, MI 48106). Mirror cells, finders, focusers.

FINISHED TELESCOPE MIRRORS (INCLUDING DIAGONALS)

COULTER OPTICAL (P.O. Box K, Idyllwild, CA 92349). Unusually large apertures, short focal lengths, and low prices. Ideal optics for large Dobsonians.

E & W OPTICAL (2420 East Hennepin Ave., Minneapolis, MN 55413). Specialists in diagonal mirrors.

ENTERPRISE OPTICS (P.O. Box 413, Dept. A, Placentia, CA 92670). Full-thickness and thin mirrors up to 18″. Refiguring service available.

A. JAEGERS (6915 Merrick Rd., Lynbrook, NY 11563). Good prices and good quality on 4″ and 6″ mirrors.

MEADE INSTRUMENTS CORP. (1675 Toronto Way, Costa Mesa, CA 92626). Same mirrors as supplied in their finished telescopes. Sizes from 6″ to 16″.

PARKS TELESCOPE COMPANY (679 Easy St., Suite B, Simi Valley, CA 93065). High prices but quality guaranteed.

SCOTT OPTICAL (4628 East Cornell, Fresno, CA 93703). Telescope mirrors 10″ and larger.

SUMMIT INSTRUMENTS (P.O. Box 152, 18803 SW 92nd Ave., Miami, FL 33157). Newtonian and Cassegrainian optics.

TELESCOPICS (P.O. Box 98, La Cañada, CA 91011). Sizes from 4″ to 20″. Newtonian and Cassegrainian optics.

EYEPIECES

CELESTRON INTERNATIONAL (2835 Columbia St., Torrance, CA 90503). Orthoscopic, Kellner, and Plossl eyepieces normally sold with quality line of Celestron telescopes.

EDMUND SCIENTIFIC (101 East Gloucester Pike, Barrington, NJ 08007). Edmund's inexpensive RKE offers the best performance per dollar in telescope eyepieces.

MEADE INSTRUMENTS CORP. (1675 Toronto Way, Costa Mesa, CA 92626). Quality import eyepieces. Erfle, Orthoscopic, Kellner in wide range of focal lengths. Same eyepieces normally sold with Meade telescopes.

TELE-VUE OPTICS, INC. (20-A Dexter Plaza, Pearl River, NY 10965). Excellent line of Plossl, Nagler, and Wide-Angle eyepieces; quite possibly the finest eyepieces being manufactured for amateur astronomy.

UNIVERSITY OPTICS (P.O. Box 1205, 2122 East Delhi Rd., Ann Arbor, MI 48106). Wide range of Plossls, Konigs, zooms and Barlows, and filters.

REFRACTOR LENSES AND KITS

ASTRO-PHYSICS (2703 Hampden Ct., Rockford, IL 61107). High quality apochromatic refractor lenses for visual and photographic applications.

A. JAEGERS (6915 Merrick Rd., Lynbrook, NY 11563). Good reputation. Finished achromats of 3″, 4″, 5″, and 6″ aperture; variety of focal ratios.

N. REMER OPTICS (P.O. Box 306 R. Southampton, PA 18966). Kits for 3″, 4″, and 6″ refractor lenses; include instructions, glass, templates, abrasives, etc. Award-winning design.

ALUMINIZING SERVICES

P. A. CLAUSING (8038 Monticello Ave., Skokie, IL 60676). Tough "beral" coatings. Good prices.

DENTON VACUUM, INC. (Cherry Hill Industrial Center, Cherry Hill, NJ 08003). Overcoated aluminum or 98%+ reflective silver coatings.

EVAPORATED METAL FILMS (701 Spencer Rd., Ithaca, NY 14850). Aluminum, enhanced aluminum, overcoated aluminum; used to dealing with amateur astronomers.

PANCRO MIRRORS, INC. (6413 San Fernando Rd., Glendale, CA 91204). Overcoated aluminum, diagonal mirrors coated free with primary.

P. A. P. COATING SERVICES (1112 Chateau Ave., Anaheim, CA 92802). Aluminum with oxide overcoat. Diagonal coated free with primary.

SUMMIT INSTRUMENTS (P.O. Box 152, 18803 SW 92nd Ave., Miami, FL 33157). Aluminum with silicon monoxide overcoat.

TELESCOPICS (P.O. Box 98, La Cañada, CA 91011). Regular overcoated aluminum and enhanced (96%) aluminum coatings.

HI-TECH ASTRONOMICAL HARDWARE AND ACCESSORIES

CELESTRON INTERNATIONAL (2835 Columbia St., Torrance, CA 90503). Accessories for astrophotography. Schmidt cameras, cold cameras, tele-compressors and extenders for Celestron telescopes.

DAYSTAR FILTERS (P.O. Box 1290, Pomona, CA 91769). Narrow-bandwidth H-alpha solar filters.

ELECTROPHYSICS CORP. (48 Spruce St., Nutley, NJ 07110). Image intensifiers, intensifier cameras, LLL-TV systems.

EDWIN HIRSCH (168 Lakeview Dr., Tomkins Cove, NY 10986). Specialist in narrow-band solar filters and viewing accessories.

LUMICON (2111 Research Dr. #5, Livermore, CA 94550). "Innovation in Astronomy"; well-designed accessory hardware for astrophotography and observing. Films, multi-layered "light-pollution" filters, guide eyepieces, film "hypering" kits, plus good and up-to-date info sheets.

THOMAS MATHIS COMPANY (2333 American Ave., Hayward, CA 94545). Observatory-quality drive gears, fork-mount castings.

OPTEC, INC. (199 Smith, Lowell, MI 49331). Equipment for stellar photometry. Solid-state, highly portable, and ready to use; with manuals.

ORION TELESCOPE CENTER (P.O. Box 1158, Santa Cruz,

CA 95061). Telescope drive correctors. Dealer for most brands.

THORN EMI GENCOM (80 Express St., Plainview, NJ 11803). Photoelectric photometers, photomultiplier tubes.

THOUSAND OAKS OPTICAL (P.O. Box 314, Wyandotte, MI 48192). Precision white-light solar filters, glass and Mylar substrates.

ROGER TUTHILL (P.O. Box 1086, Mountainside, NJ 07092). Mylar solar filters, narrow-band and prominence filters, digital-position readouts.

SURPLUS

AMERICAN SCIENCE CENTER (5700 Northwest Highway, Chicago, IL 60646; mail-order outlet, Jerryco Inc., 601 Linden Pl., Evanston, IL 60202). A wild mix of surplus stuff from optics to pickled frogs to stepper motors, plus the complete Edmund Scientific line.

T. R. INC. (P.O. Box 65, Mooers, NY 12958). High-grade, high-tech surplus optics and instruments. Send for a catalog.

TOOL AND HARDWARE SUPPLIERS

BROOKSTONE COMPANY (127 Vose Farm Rd., Peterborough, NH 03548; Tel: 603-924-9511). "Hard-to-Find Tools and Other Things"—a wonderful assortment of fine woodworking tools, boxes, bins, and gadgets for the home workshop. Good quality, prompt delivery.

J. CHEAPS & SONS (P.O. Box 7199, Warrensville, OH 44128). Tools and gadgets for the home shop. Carries threaded insert hardware, tee nuts, and right-angle joint connectors.

GARRETT WADE CO. (161 Avenue of the Americas, New York, NY 10013). "Our catalog doesn't just sell you things. It teaches you things." Catalog costs $3 (1982), but is worth every penny. Tells how to choose tools, and includes lots of photos and "honest specifications."

JENSEN TOOLS & ALLOYS (1230 South Priest Dr., Tempe, AZ 85281). A complete line of tools and tool kits; especially suitable for model-makers and traveling electronics repairmen.

LEICHTUNG, INC. (4944 Commerce Parkway, Cleveland, OH 44128). Quality tools for woodworking. Carries threaded insert hardware, tee nuts, and right-angle joint connectors.

SEARS, ROEBUCK AND CO. A huge retailer of hardware and

tools with stores all across the United States. Always buy their best-grade tools.

SHOPSMITH INC. (6640 Poe Ave., Dayton, OH 45414). Every kind of woodworking tool for home workshops.

SMALL PARTS, INC. (6901 NE Third Ave., Miami, FL 33138; Tel: 305-751-0856). "A Selection of High-Quality Small Mechnical Parts and Components for Research and Development." All sorts of specialized items, perfect if you're the telescope-maker with a machine-shop bent. Fast delivery.

Appendix C
Books and Other Information Resources

CONSTELLATIONS AND STAR LORE

Allen, Richard Hinckley, *Star Names—Their Lore and Meaning*, Dover, New York, NY, 1963; republication of *Star Names and Their Meanings*, publ., G.E. Stechert, 1899.

Invaluable while learning the stars. Helps you understand what the names mean and cultural influences of and on astronomy.

Menzel, Donald H., and Pasachoff, Jay M., *A Field Guide to the Stars and Planets*, second revised edition, Houghton-Mifflin, Boston, MA, 1983.

Part of the Peterson Field Guide Series. Contains monthly star maps, rather cramped star charts by Wil Tirion, and lunar atlas. Not an especially attractive or appealing book, but widely available.

Ottewell, Guy, *Astronomical Calendar*, Guy Ottewell, Dept. of Physics, Furman Univ., Greenville, SC; self-published annual since 1974.

A 52-page, beginner-oriented, month-by-month guide to the skies. Useful data on planetary phenomena, finding asteroids, watching meteor showers, etc.

Philips' Planisphere, George Philip and Son, Ltd., London,

"Showing the principal stars visible for every hour in the year." Clearly defined constellations printed on heavy waterproof plastic circles. Probably the best.

Raymo, Chet, *365 Starry Nights, an introduction to astronomy for every night of the year*, Prentice-Hall, Englewood Cliffs, NJ, 1982.

A delightfully illustrated night-by-night "go out and look" book. Introduction to the major constellations and interesting facts about stars and nebulae.

Ronan, Colin A., *The Practical Astronomer*, Macmillan, New York, NY, 1981.

A good general introduction to astronomy, well-illustrated, with lots of hints about projects the reader could undertake.

OBSERVING GUIDES

Burnham, Robert, Jr., *Burnham's Celestial Handbook*, in three volumes, Dover, New York, NY, 1978; expanded and updated republication of the work originally published by Celestial Handbook Publications, Flagstaff, AZ, 1966.

The one book no serious observer can be without. Totaling over 2000 pages, it presents, for each constellation, lists of double and multiple stars, clusters, nebulae, and galaxies. Extended descriptions of interesting objects, many photographs; includes 74 pages on Orion.

Fulton, Ken, *The Light-Hearted Astronomer*, Astromedia, Milwaukee, WI, 1984.

A gentle and humorous guide to the art of *being* an amateur astronomer. Fulton strikes all the emotional chords in his journey through astronomy's jungle. Worth reading once a year.

Holt, Terry, *The Universe Next Door*, Charles Scribner's Sons, New York, NY, 1985.

The first introductory guide to combine descriptive astronomy with tips and techniques for observing and telescope use.

Howard, Neal E., *The Telescope Handbook and Star Atlas*, Thomas Y. Crowell, New York, NY, 1975.

An introduction to the sky and to observing. Transparent overlay maps of telescopic objects not as effective as they should be.

Mallas, John H., and Kreimer, Evered, *The Messier Album*, Sky Publishing Corp., Cambridge, MA, 1978.

Each Messier object gets a page or more, with finder map, photograph, and drawing. Excellent for beginner. Has interesting historical introduction.

Minnaert, Marcel, *Light and Colour in the Open Air*, Dover, New York, NY, 1954; republication of same title, G. Bell & Sons.

Light, the essence of astronomical observation, described and explored. Information on intensity, color, afterimages, contrast phenomena and light and color as seen in the sky, rainbows, halos.

Moirden, James, *The Amateur Astronomer's Handbook*, 3rd ed., Harper & Row, New York, NY, 1983.

Not to be confused with Sidgewick's book with *nearly* the same title, this book covers the broad scope of hobby astronomy in a clear, readable fashion. A must for serious observers.

ASTRONOMICAL PERIODICALS

ASTRONOMY, AstroMedia, 625 E. St. Paul Ave., Milwaukee, WI 53202

Monthly, 96 pages, $18/year. Colorful, lavishly illustrated feature articles in nontechnical language, monthly departments on equipment, observing, photographing, what to see; monthly star chart; monthly constellation close-up; naked-eye and telescopic star and planetary observing columns. Available by subscription, on newsstands, and in most public libraries.

Deep Sky, AstroMedia, 625 E. St. Paul Ave., Milwaukee, WI 53202

Quarterly, 40 pages, $12/year. Devoted to enthusiastic observers of deep-sky objects; considerable emphasis on astrophotography.

Odyssey, AstroMedia, 625 E. St. Paul Ave., Milwaukee, WI 53202

Monthly, 40 pages, $12/year. Space and science magazine for bright youngsters in the 8 to 14 age range; monthly star map, science project, NASA news.

Sky & Telescope, Sky Publishing Corp., 49 Bay State Rd., Cambridge, MA 02238

Monthly, 96 pages, $18/year. Semi-technical journal for advanced amateur astronomers; historical articles, features, current research news, valuable column on telescope building titled "Gleanings for ATMs," monthly star chart, observing column titled "Deep Sky Wonders."

Telescope Making, AstroMedia, 625 E. St. Paul Ave., Milwaukee, WI 53202

Quarterly, 48 pages, $12/year. Small but influential magazine with an emphasis on innovative ideas; each issue carries six to ten articles by telescope-builders, lots of photographs, convention and meeting reports.

Note: Newsletters, publications, and proceedings are listed with the organizations publishing them; see "Astronomical Organizations for Amateur Astronomers."

BOOKS ON TELESCOPES, OBSERVATORIES, AND THEIR HISTORY

Bell, Louis, *The Telescope*, Dover, New York, NY, 1981; republication of the original by McGraw-Hill, 1922.

Still as useful as the day it was written, "Bell's Book of the Telescope" is a valuable and illuminating discussion of telescopes, how they work, their parts and accessories, taking care of them, and using them. Especially valuable on refractors.

King, Henry C., *The History of the Telescope*, Dover, New York, NY, 1981; republication of the original by Charles Griffin, London, 1955.

A rich source of information—wonderful to read over and over again. Best for eras before 1920, and for European and British makers. Many old illustrations reproduced.

Learner, Richard, *Astronomy Through the Telescope*, Van Nostrand Reinhold, New York, NY, 1981.

Essentially historic approach. Provides a broad overview of telescopes and their role in astronomical research. Not as detailed as King, but much more graphic.

Sidgewick, J.B., *Amateur Astronomer's Handbook*, 3rd ed.: Dover, New York, NY, 1980; republication of the original 3rd ed. by Faber and Faber, London, 1971. 4th ed.: revised by James Muirden; Enslow, Hillside, NJ, 1980.

A great deal of useful information on telescopes, telescope mountings, and their accessories.

Warner, Deborah J., *Alvan Clarke & Sons, Artists in Optics*, Smithsonian Institution Press, Washington, DC, 1968.

How the premier opticians in America did their work during the mid-nineteenth century.

Willard, Berton C., *Russell W. Porter: Arctic Explorer, Artist, Telescope Maker*, Bond Wheelwright Co., 1976; order from Willmann-Bell, Inc.

A good read. Porter, the "patron saint" of amateur telescope-making in America, founded Stellafane, contributed heavily to the Amateur Telescope Making books, and was an important contributor to the 200″ Palomar telescope project.

Woodbury, David O., *The Glass Giant of Palomar*, Dodd, Mead & Co., New York, NY, 1939, 1952, rev. ed., 1970.

Inspiring reading while you're building one of your own. Written when science was still a "good thing" and very popular in America.

BOOKS ON TELESCOPE BUILDING

Ingalls, Albert G., ed., *Amateur Telescope Making, Book One*, Scientific American, Inc., New York, NY, 1935, 1941, 1945.

Made up from a large number of contributions, this wonderful but somewhat disorganized book is *the* classic how-to-do-it source for basic techniques. Excellent material, but somewhat dated.

———, ed., *Amateur Telescope Making Advanced*, Scientific American, Inc., New York, NY, 1937, 1944, 1946.

Advanced mirror-making techniques; good material on refractors, metal casting, and machining. Better organized than *Book One*.

———, ed., *Amateur Telescope Making—Book Three*, Scientific American, Inc., New York, NY, 1953.

Lens design, H-alpha filters, Schmidt cameras, optical testing dominate this volume, the last of the ATM series.

How to Build Your Own Observatory, AstroMedia Corp., Milwaukee, WI, 1981.; revised and expanded 2nd ed., 1985.

Reprints from *Telescope Making* magazine. Ten extremely useful articles including simple shelters, converted yard shed buildings, garage-top domes, lift-off and roll-off roofs, fiberglassed domes. 66 pages.

Howard, Neale E., *Standard Handbook for Telescope Making*, rev. ed., Harper & Row, New York, NY, 1984.

Assumes you are making an 8″ f/7 Newtonian. Approximately 50% devoted to grinding the mirror. Excellent section on working with fiberglass, weak on mountings.

Kestner, Robert, et al., *How to Build a Dobsonian Telescope*, AstroMedia Corp., Milwaukee, WI, 1980; revised and expanded 2nd ed., 1982.

Reprints from *Telescope Making* magazine. Includes mounting basics, friction, use of setting circles, Poncet platform tracking, interview with John Dobson. 52 pages.

Taylor, H. Dennis, *The Adjustment and Testing of Telescope Objectives*, 5th ed., Adam Hilger, Ltd., Bristol, 1983; republication of the 4th and earlier ed., Messrs. T. Cooke & Sons, 1891, 1896, 1921, 1946.

112 pages of refractor lore, stresses the importance of proper collimation, evaluation of objective quality, and extrafocal star testing. Available in U.S. from Heyden & Son, Inc., Philadelphia.

Texereau, Jean, *How to Make a Telescope*, 2nd English ed. Willmann-Bell, Richmond, VA, 1984; translated and adapted from "La contruction du telescope d'amateur," 2nd ed., originally published in *l'Astronomie*, 1951.

The best book on mirror-making, oriented toward the optician. Texereau is a perfectionist, so he leads you, step by step, toward the goal of completing a near-perfect mirror. The book assumes you are making an 8″ f/6 Newtonian, although it is general enough that you can use it to undertake almost any mirror-grinding project. Good section on yoke and cross-axis mounts. Appendices include listing of significant periodical literature from 1946 on.

Thompson, Allyn J., *Making Your Own Telescope*, Sky Publishing Corp., Cambridge, MA, 1947.

The basic prescription for the standard amateur 6″ f/8 Newtonian of the 50s and 60s. Assumes some access to a machine shop. Dated but still useful.

Trueblood, Mark, and Genet, Russell, *Microcomputer Control of Telescopes*, Willmann-Bell, Richmond, VA, 1985.

An exhaustive treatise on fully automating amateur-size telescopes. Includes both hardware and software information.

ARTICLES ON TELESCOPE BUILDING

Berry, Richard, "Astronomy's Neglected Child—The Long Refractor," *Sky and Telescope*, Vol. 51, No. 2, p. 130; Feb., 1976.

The construction and performance of a 60mm-aperture

simple-lens refractor, slung from a pole with ropes. Concludes the old simple lens refractors weren't so bad. Historical interest.

———, "Newtonian Telescopes," *Telescope Making #9*, Fall 1980.

Formulae, graphs, and charts for the imaging-forming properties and aberrations of the Newtonian optical system.

———, "Yoke and Cross-Axis Mountings Made Easy," *Telescope Making #11*, Spring 1981.

How to build these successfully from wood, why they work well, lots of photos.

Buchroeder, Richard A., "Cassegrain Optical Systems," *Telescope Making #1*, Fall 1978 (see also corrections in TM#2 p. 32, and TM#3, p. 3).

Formulae for calculating classical, Ritchey-Chretien, and Dall-Kirkham primaries and secondaries.

Cox, Robert E., "Improving Telescope Performance— Part 1, Better Mirror Cells," *Telescope Making #6*, Winter 1979/80; and "Part 2, Secondary Mirrors and Spiders," *Telescope Making #7*, Spring 1980.

How to modify commercial mirror cells for better performance.

Christen, Roland, "An Apochromatic Triplet Objective," *Sky and Telescope*, Vol. 62, No. 4, p. 376; October 1981.

Useful tips on fabricating a triplet objective. Information on glass sources, available materials, mounting the lens.

Dey, Tom, "Dobsonian Developments," *Telescope Making #15*, Spring 1982.

Explores the good and bad points of the Dobsonian design. Gives techniques for aligning an alt-azimuth mounting with setting circles.

Hamberg, Ivar, "An Extremely Portable 17.5″ Dobsonian," *Telescope Making #17*, Fall 1982.

This two-page article with just three pictures should inspire anyone intending to build a large-aperture, portable telescope.

Kestner, Robert, "Grinding, Polishing and Figuring Large Thin Mirrors, Part 1—Grinding," *Telescope Making #12*, Summer 1981; "Part 2—Polishing," *Telescope Making #13*, Fall 1981; "Part 3—Figuring," *Telescope Making #16*, Summer 1982.

The essential element in large-aperture Dobsonian telescopes is the mirror. These articles tell how to make a mirror up to 25″ in diameter on glass only 1.63″ thick. Lots of practical detail on every operation. Author assumes reader has some previous mirror-making experience.

Millies-Lacroix, Adrien, "A Graphical Approach to the Foucault Test," *Sky and Telescope*, Vol. 51, No. 2, p. 127; see also Robert P. Miller, "Stalking the Wild Paraboloid," *Telescope Making #6*, Summer 1980.

One of the first English-language discussions of the envelope method for reducing Foucault test measurements graphically. A later discussion by a user.

Stolzmann, David, "Newtonian Aberrations," *Telescope Making #12*, Summer 1981.

Formulae for computing third-order aberrations of the Newtonian telescope with a small computer. Worked-out spot diagrams included.

van Venrooij, M.A.M., "Contrast Rendition and Resolving Power in Telescopes Used Visually," *Telescope Making #16*, Summer 1982.

Why unobstructed optical systems are better, and how the obstruction decreases image contrast.

Wilson, Jim, "How to Make a Lens," *Telescope Making #3*, Spring 1979.

A thorough description of lens fabrication, from making templates and tools to edging the finished piece. Essential reading if you're thinking of making a refractor objective although the emphasis is on smaller lenses.

BOOKS ON OPTICS

Conrady, A. E., *Applied Optics and Optical Design*, Dover, New York, NY, 1957; republication of the original by Oxford, 1929.

Two volumes. Old-fashioned and superseded by Kingslake, but still worth having on hand if you're interested in optical design.

DeVany, Arthur S., *Master Optical Techniques*, John Wiley & Sons, New York, NY, 1981.

Not for the beginner. Covers a wide range of workshop problems likely to be encountered by a professional optician. Excellent for an advanced amateur optician.

Hecht, Eugene, and Zajac, Alfred, *Optics*, Addison-Wesley, Reading, MA, 1974.

The clearest and most beautifully illustrated optics text I know. Not much help to the practical optician, but you'll really understand light if you can work your way through this book. Requires math.

Kingslake, Rudolf, *Lens Design Fundamentals*, Academic Press, New York, NY, 1978.

For the designer. Very clear development; equations prepared for use with calculators and computers rather than log and trig tables. Valuable sections on calculating achromats, apochromats, catadioptric systems, and eyepieces.

Rose, Albert, *Vision: Human and Electronic*, Plenum Press, New York, NY, 1973.

> No part of the telescope is as important as the detector on the back end. This book provides a readable theoretical understanding of vision.

Smith, Warren J., *Modern Optical Engineering*, McGraw-Hill, New York, NY, 1966.

> One of the best. Covers the whole field from materials to designing to tolerancing finished parts. Excellent sections on ray tracing. Requires math.

STAR MAPS AND ATLASES

Norton, Arthur P., *Norton's Star Atlas and Telescopic Handbook*, Gall and Inglis, London, 1910; 17 subsequent editions; republished in the USA by Sky Publishing Corp., Cambridge, MA, 1966.

> Eight star maps covering the whole sky to magnitude 6.5; 8000 stars, nebulae, and clusters, epoch 1950. Descriptive lists of objects suitable for small telescopes. Introductory section on telescopes and astronomy. A very pleasant atlas for a beginner.

Scoville, Charles, and Christos Popadopoulos, *True Visual Magnitude Photographic Star Atlas*, in three volumes, Pergammon, 1979 and 1980.

> 456 15″x15½″ charts, made in yellow light, show black stars on a white background to 13.5 mag. Expensive (about $500). Not for the beginner.

Tirion, Wil, *Sky Atlas 2000.0*, deluxe ed., Cambridge University Press, Cambridge (England), 1981.

> 26 large charts cover the whole sky; 43,000 stars to 8.0 mag., 2500 deep-sky objects, color-coded. A clear, uncluttered star atlas suitable for use with a first telescope and good enough to satisfy the user for many years. Also available in smaller monochrome "field" and "desk" editions.

Vehrenberg, Hans, *Atlas of Deep Sky Splendors*, Sky Publishing Corp., Cambridge, MA, 1967; English-language republication of *Mein Messier Buch*, Treugesell-Verlag, Dusseldorf, 1966.

> A beautiful book consisting of photographs of Messier and NGC objects at a uniform scale of 60mm/degree taken with the author's Schmidt camerw stars to roughly magnitude 14 at a scale of 30mm/degree.

———, *Photographic Star Altas*, Treugesell Verlag, Dusseldorf, 1972.

> Shows stars to magnitude 13 at a scale of 15mm/degree; 464 maps of line reproductions of star photographs. Often used as background maps for plotting asteroid paths.

STAR AND DEEP-SKY OBJECTS CATALOGS

Hirshfeld, Alan, and Roger Sinnott, *Sky Catalog 2000.0, Volume 1: Stars to Magnitude 8.0*, Cambridge University Press, Cambridge (England), 1982.

> Looks like a big phone book. 75 stars per page, 604 pages; 45,269 stars listed in order of RA with positions, proper motions, magnitudes, spectral class, distance. Essentially complete to the limit of this catalog.

———, *Sky Catalog 2000.0, Volume 2: Double Stars, Variable Stars, and Non-Stellar Objects*, Cambridge University Press, Cambridge (England), 1985.

> A wealth of information on over 21,000 deep-sky objects. 156 pages devoted to doubles. 3100 galaxies, planetary nebulae, globular clusters, and quasars. 385 pages.

Hoffleit, Dorrit, with Jaschek, Carlos, *Catalog of Bright Stars*, 4th ed., Yale University, New Haven, CT, 1982.

> Nearly 400 pages of stellar data, magnitude, spectral, proper motion data on 9091 stars to magnitude 6.5. 75 pages of supplementary remarks, cross-referenced to other catalogs.

Kukarkin, B. V., et al., *General Catalogue of Variable Stars*, 3 Vols., USSR Academy of Sciences, Moscow, 1969, 1975.

> The standard reference for the variable-star observer. Lists all known and suspected variable stars.

Sandage, Alan, and Tammann, G. A., *A Revised Shapley-Ames Catalog of Bright Galaxies*, Carnegie Institute of Washington, Washington, DC, 1981.

> Data on the 1246 brightest galaxies. A must for the serious deep-sky observer who wants to know what he's looking at. Not for the beginner.

Sulentic, Jack W., and Tifft, William G., *The Revised New General Catalogue of Nonstellar Astronomical Objects*, University of Arizona Press, Tucson, AZ, 1973.

> A 365-page computer listing, using cryptic letter codes, of the properties of over 8000 objects in the NGC catalog. Not for the beginner.

MAKING A SCIENTIFIC CONTRIBUTION TO ASTRONOMY

Couteau, Paul, *Observing Visual Double Stars,* The MIT Press, Cambridge, MA, 1981; translation by Alan Batten of "L'observation des étoiles doubles visuelles," Flammarion, Paris, 1978.

A detailed and thorough work, both practical and complete. Sections on computing orbits and star masses. Useful table of 744 double stars.

Hall, Douglas S., and Genet, Russell M., *Photoelectric Photometry of Variable Stars,* Interitable for use with a first telescope and good enough to satisfy the user for many years. Also available in smaller monochrome 'national-Amateur Professional Photoelectric Photometry, Fairborn, OH, 1982.

A practical hands-on guide with contributions by a dozen members of IAPPP. Lots of examples of amateur observatories. Sections on taking and reducing photoelectric data.

Henden, Arne A., and Kaitchuck, Ronald H., *Astronomical Photometry*, Van Nostand Reinhold Co., New York, NY, 1982.

A very clear and well-organized account of what photometry is, how it works, the equipment and equations necessary for its pursuit. Useful tables and appendices.

Povenmire, Harold R., *Fireballs, Meteors, and Meteorites,* JSB Enterprises, FL, 1980.

The "how-to" of meteor observing. American Meteor Society supports these observations.

————, *Graze Observer's Handbook*, Vantage Press, New York, NY, 1975.

Techniques for observing grazing occultations of stars by the moon, written by an enthusiastic graze observer. IOTA supports this type of observation.

Roth, Gunter D., *Astronomy: A Handbook*, Springer-Verlag, New York, NY 1975; translated by Auther Beer from *Handbuch fuer Sternfreunde*, 2nd ed., Springer-Verlag, 1967.

A compendium of valuable information, very rigorous and technical, loaded with equations, tables, and graphs.

Sidgewick, J. B., *Observational Astronomy for Amateurs*, Dover, New York, NY, 1980, republication of the 3rd ed., Faber and Faber, London, 1971. 4th ed.: rev. by James Muirden, Enslow, Hillside, NJ, 1982.

Covers the range of observations amateur astronomers could have been expected to undertake in the 50s or 60s. Assumes data collection through the BAA.

United States Naval Observatory, *Astronomical Almanac*, prior to 1981 called *The American Ephemeris and Nautical Almanac*, U.S. Gov't Printing Office, Washington, DC, published annually.

The prime source of data on planetary positions and physical data, eclipse times, occultations, upcoming events for each year.

ASTRONOMICAL ORGANIZATIONS FOR AMATEUR ASTRONOMERS

AMERICAN ASSOCIATION OF VARIABLE STAR OBSERVERS, 187 Concord Ave., Cambridge, MA 02138; Contact: Dr. Janet Mattei, 617-354-0484. Publishes *Journal of the AAVSO.*

AMERICAN METEOR SOCIETY, 4238 Springwood Rd., Jacksonville, FL 32207; Contact: Harold Povenmire. Publishes *Meteor News.*

ASSOCIATION OF LUNAR AND PLANETARY OBSERVERS, Box 3AZ, University Park, NM 88003; Contact: Walter Haas. Publishes *The Strolling Astronomer.*

ASTRONOMICAL LEAGUE, P.O. Box 12821, Tucson, AZ 85723; Contact: Don Archer. Publishes *The Reflector.*

ASTRONOMICAL SOCIETY OF THE PACIFIC, 1290 24th Ave., San Francisco, CA 94122; Contact: Andrew Fraknoi, 415-661-8660. Publishes *Mercury* and *Publications of the ASP.*

ASTRONOMICAL SOCIETY OF NEW SOUTH WALES (Australia), P.O. Box 208, Eastwood, New South Wales 2122.

ASTRONOMICAL SOCIETY OF SOUTH AUSTRALIA, P.O. Box 199, Adelaide, South Australia 5001.

ASTRONOMICAL SOCIETY OF SOUTH AFRICA, c/o South African Astronomical Observatory, P.O. Box 9, Observatory 7935, Capetown, South Africa.

INTERNATIONAL AMATEUR-PROFESSIONAL PHOTOELECTRIC PHOTOMETRY, 629 North 30th St., Phoenix, AZ 85008; Contact: Russell Genet. Publishes *Proceedings of the IAPPP.*

INTERNATIONAL HALLEY WATCH, AMATEUR OBSERVATION NET, MS T-1166, Jet Propulsion Laboratory, 4800 Oak Grove Rd., Pasadena, CA 91109; Contact: Steve Edberg. Publishes a newsletter.

BRITISH ASTRONOMICAL ASSOCIATION (amateur) and ROYAL ASTRONOMICAL SOCIETY (professional), Burlington House, Picadilly, London, W1V ONL, England. BAA publishes *Journal of the BAA, Circulars,* and, annually, *Handbook.*

INTERNATIONAL OCCULTATION TIMING ASSOCIATION, P.O. Box 596, Tinley Park, IL 60477.

JUNIOR ASTRONOMICAL SOCIETY (British), c/o Mr. V. L.

Tibbott, 58 Vaughan Gardens, Ilford, Essex IG1 3PD.
Publishes *Hermes*.
ROYAL ASTRONOMICAL SOCIETY OF CANADA, 124 Merton
St., Toronto, Ontario M4S 2Z2, Canada. Centers all
across Canada. Publishes *Journal of the RASC* and
annual *Observer's Handbook*.
ROYAL ASTRONOMICAL SOCIETY OF NEW ZEALAND, P.O.
Box 3181, Wellington, C1.
WESTERN AMATEUR ASTRONOMERS, 1110 Petaluma Hill
Rd., Santa Rosa, CA 95404; Contact: Frank A. Miller.

MAJOR ANNUAL
TELESCOPE CONFERENCES

MIDWEST ASTROFEST: Held near Kankakee, IL, in late
September. Camping or bunk beds available in dor-
mitories. About 150 attendees.
RIVERSIDE TELESCOPE MAKERS CONFERENCE: Held near Big
Bear City, CA, on Memorial Day weekend, elevation
7500 feet. Commercial exhibitors. Camping or bunk
beds available in dormitories. About 850 attendees.
STELLAFANE: Held annually on a new-moon weekend in
late July or early August on Breezy Hill, near Springfield,
VT. No commercial exhibitors. Limited primitive
camping at site, other campgrounds and motels nearby.
About 1000 attendees.
THE TEXAS STAR PARTY: Held in West Texas, where the
sky is really dark, in late May or early June. Commercial
exhibitors. Camping or bunk beds available in dormitories, about 250 attendees.

Watch for specific annual information in the astronomical monthlies.

BOOKS ABOUT ASTRONOMY
AND SPACE

Berendzen, Richard; Hart, Richard; and Seeley, Daniel,
Man Discovers the Galaxies, Science History Pub-
lications, New York, NY, 1976.
Bok, Bart J.; and Bok, Pricilla, *The Milky Way*, 5th ed.,
Harvard University Press, Cambridge, MA, 1981.
Chapman, Clark, *Planets of Rock and Ice*, Charles
Scribner's Sons, New York, NY, 1982.
Davies, Merton E., chief ed., *Atlas of Mercury*, NASA
SP-423, Washington, DC, 1978.
Davies, P. C. W., *The Accidental Universe*, Cambridge
University Press, Cambridge (England), 1982.

Ferris, Timothy, *The Red Limit*, William Morrow & Co.,
New York, NY, 1977.
———, *Galaxies*, Sierra Club Books, San Francisco, CA,
1980; reprinted in smaller format by Stewart, Tabori
and Chang, New York, NY, 1982.
Hartmann, William K., *Moons and Planets*, 2nd ed.,
Wadsworth, Belmont, CA, 1972, 1983.
Mailer, Norman, *Of A Fire On The Moon*, Little, Brown
and Co., Boston, MA, 1969.
Malin, David; and Murden, Paul, *Colours of the Stars*,
Cambridge University Press, Cambridge (England),
1984.
Mitton, Simon, ed., *Cambridge Encyclopedia of Astron-
omy*, Crown, New York, NY, 1977.
Moore, Patrick, *The Unfolding Universe*, Crown, New
York, NY, 1982.
Morrison, David, *Voyages to Saturn*, NASA SP-451,
Washington, DC, 1982.
———, and Samz, Jane, *Voyage to Jupiter*, NASA SP-439,
Washington, DC, 1980.
Muirden, Paul, et al., *Catalogue of the Universe*, Crown,
New York, NY, 1980.
Pasachoff, Jay, *University Astronomy*, Saunders, Phila-
delphia, PA, 1978.
Peltier, Leslie C., *Starlight Nights: The Adventures
of a Star-Gazer*, Sky Publishing Corp., Cambridge,
MA, 1965.
Sagan, Carl, *Cosmos*, Random House, New York, NY,
1980.
Silk, Joseph, *The Big Bang: The Creation and Evolution
of the Universe*, W. H. Freeman, San Francisco, CA,
1980.
Shipman, Harry L., *Black Holes, Quasars, and the
Universe*, 2nd ed., Houghton-Mifflin & Co., Boston,
MA, 1980.
Shu, Frank H., *The Physical Universe: an Introduction to
Astronomy*, University Science Books, Mill Valley, CA,
1982.
Tombaugh, Clyde, and Moore, Patrick, *Out of Darkness:
The Planet Pluto*, Stackpole, Harrisburg, PA, 1980.
Viking Lander Imaging Team, *The Martian Landscape*,
NASA SP-425, Washington, DC, 1978.
Whitney, Charles A., *The Discovery of Our Galaxy*,
Alfred A. Knopf, New York, NY, 1971.
Wolfe, Tom, *The Right Stuff*, Farrar, Straus, Giroux, New
York, NY, 1979.

Index